ANIMAL
TRACKS
of
BRITISH
COLUMBIA

D0001888

Ian Sheldon & Tamara Hartson

LONE
PINE

THE PUBLISHER: LONE PINE PUBLISHING

1808 B Street NW, Suite 110	10145 - 81 Avenue
Auburn, WA 98001	Edmonton, AB T6E 1W9
USA	Canada

Lone Pine Publishing website: http://www.lonepinepublishing.com

Canadian Cataloguing in Publication Data

Sheldon, Ian
  Animal tracks of British Columbia

  Includes bibliographical references and index.
  ISBN-13: 978-1-55105-223-6
  ISBN-10: 1-55105-223-7

  1. Animal tracks—British Columbia—Identification.
I. Hartson, Tamara II. Title.

QL768.S523 1999   591.47'9   C99-910531-0

Editorial Director: Nancy Foulds
Editor: Volker Bodegom
Proofreader: Lee Craig
Production Manager: Jody Reekie
Design, layout and production: Volker Bodegom, Michelle Bynoe
Cartography: Volker Bodegom
Animal illustrations: Gary Ross, Horst Krause, Ian Sheldon
Track illustrations: Ian Sheldon
Cover illustration: Mountain Lion by Gary Ross
Scanning: Elite Lithographers Ltd.

We acknowledge the financial support of the Government of Canada through
the Book Publishing Industry Development Program (BPIDP) for our publishing activities.

PC: 06   Canadä

CONTENTS

INTRODUCTION

If you have ever spent time with an experienced tracker, or perhaps a veteran hunter, then you know just how much there is to learn about the subject of tracking and just how exciting the challenge of tracking animals can be. Maybe you think that tracking is no fun, because all you get to see are the animal's prints. What about the animal itself—is that not much more exciting? Well, for most of us who do not spend a great deal of time in the beautiful wilderness of British Columbia, the chances of seeing the secretive Mountain Lion or the fun-loving River Otter are slim. The closest that we may ever get to some animals will be through their tracks, and they can inspire a very intimate experience. Remember, you are following in the footsteps of the unseen—animals that are in pursuit of prey, or perhaps being pursued as prey.

This book offers an introduction to the complex world of tracking animals. Sometimes tracking is easy. At other times, it is an incredible challenge that leaves you wondering just what animal left those unusual tracks. Take this book into the field with you, and it can provide some help with the first steps to identification. Animal tracks and trails are this book's focus; you will learn to recognize subtle differences for both. There are, of course, many additional signs to consider, such as scat and food caches, all of which help you to understand the animal that you are tracking.

Remember, it takes many years to become an expert tracker. Tracking is one of those skills that grows with you as you acquire new knowledge in new situations. Most importantly, you will have an intimate experience with nature. You will learn the secrets of the seldom seen. The more you discover, the more you will want to know, and by developing a good understanding of tracking, you will gain an excellent appreciation of the intricacies and delights of our marvellous natural world.

How to Use this Book

Most importantly, take this book into the field with you! Relying on your memory is not an adequate way to identify tracks. Track identification has to be done in the field, or with detailed sketches and notes that you can take home. Much of the process of identification is circumstantial, so you will have much more success when standing beside the track.

This book is laid out to be easy to use. There is a quick reference appendix to the tracks of all the animals illustrated in this book starting on p. 140. This appendix is a fast way to familiarize yourself with certain tracks and the content of this book, and it guides you to the more informative descriptions of each animal and its track.

Each animal's description is illustrated with the appropriate footprints and the trail patterns that it usually leaves. Though these illustrations are not exhaustive, they do show the tracks or groups of prints that you will most likely see. You will find a list of dimensions for the tracks, giving the general range, but there will always be extremes, just as there are with people who have unusually small or large feet. Under the category 'Size' (of animal), the 'greater-than' sign (>) is used when the size difference between the sexes is pronounced.

If you think that you may have identified a track, check the 'Similar Species' section for that animal. This section is designed to help you confirm your conclusions by pointing out other animals that leave similar tracks and showing you ways to distinguish among them.

As you read this book, you will notice an abundance of words such as 'often,' 'mostly' and 'usually.' Unfortunately, tracking will never be an exact science; we cannot expect animals to conform to our expectations, so be prepared for the unpredictable.

Tips on Tracking

As you flip through this guide, you will notice clear, well-formed prints. Do not be deceived! It is a rare track that will ever show so clearly. For a good, clear print, the perfect conditions are slightly wet, shallow snow that is not melting, or slightly soft mud that is not actually wet. Needless to say, these conditions can be rare—most often you will be dealing with incomplete or faint prints, where you cannot really be sure of the number of toes.

Should you find yourself looking at a clear print, then the job of identification is much easier. There are a number of key features to look for: measure the length and width of the print, count the number of toes, check for claw marks and note how far away they are from the body of the print, and look for a heel. Keep in mind other more subtle features, such as the spacing between the toes, whether or not the toes are parallel, and whether the fur on the sole of the foot has made the print less clear.

When you are faced with the challenge of identifying an unclear print—or even if you think that you have made a successful identification from one print alone—look beyond the single footprint and search out others. Do not rely on the dimensions of one print alone, but collect measurements from several prints to get an average. Even the prints within one trail can show a lot of variation.

Try to determine which is the fore print and which is the hind, and remember that many animals are built very differently from humans, having larger forefeet than hind feet. Sometimes the prints will overlap, or they can be directly on top of one another in a direct register. For some animals, the fore and hind prints are pretty much the same size.

Check out the pattern that the tracks make together in the trail, and follow the trail for as many paces as is necessary for you to become familiar with the pattern. Patterns are very important, and can be the distinguishing feature between different animals with otherwise similar tracks.

Follow the trail for some distance, because it may give you some vital clues. For example, the trail may lead you to a tree, indicating that the animal is a climber—or it may lead down into a burrow. This part of tracking can be the most rewarding, because you are following the life of the animal as it hunts, runs, walks, jumps, feeds or tries to escape a predator.

Take into consideration the habitat. Sometimes very similar species can be distinguished only by their locations— one might be found on riverbanks, whereas another might be encountered only in the dense forest.

Think about your geographical location, too. Some animals have a limited range—perhaps they are found only in remote parts of British Columbia. This consideration can rule out some species and help you with your identification.

Remember that every animal will at some point leave a print or trail that looks just like the print or trail of a completely different animal!

Lastly, keep in mind that if you track quietly, you might catch up with the maker of the prints.

Terms and Measurements

Some of the terms used in tracking can be rather confusing, and they often depend on personal interpretation. For example, what comes to your mind if you see the word 'hopping'? Perhaps you see a person hopping about on one leg—or perhaps you see a rabbit hopping through the countryside. Clearly, one person's perception of motion can be very different from another's. Some useful terms are explained below, to clarify what is meant in this book and, where appropriate, how the measurements given fit in with each term.

Red Squirrel

The following terms are sometimes used loosely and interchangeably—for example, a rabbit might be described as a 'hopper' and a squirrel as a 'bounder,' yet both have the same gait and leave the same pattern of prints in the same sequence.

Ambling: Fast, rolling walking.

Bounding: A gait of four-legged animals in which the two hind feet land simultaneously, usually registering in front of the forefeet. Common in rodents and the rabbit family. 'Hopping' or 'jumping' can often be substituted.

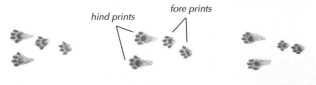

hind prints   fore prints

Gait: An animal's gait describes how it is moving at some point in time, and results in observable trail characteristics.

Galloping: A gait used by animals with four even-length legs, such as dogs, moving at high speed, hind feet registering in front of forefeet.

hind prints   *fore prints*   *gallop group*

Hopping: Similar to bounding. With four-legged animals, usually indicated by tight clusters of prints, fore prints set between and behind the hind prints. A bird hopping on two feet creates a series of paired tracks along its trail.

Loping: Like galloping, but slower, with each foot falling independently and leaving a trail pattern that consists of groups of tracks in the sequence fore-hind-fore-hind, usually roughly in a line.

Mustelids (weasel family) often use ***2x2 loping***, in which the hind feet register directly on the fore prints. The resulting pattern has angled, paired tracks.

Running: Like galloping, but applied generally to animals moving at high speed. Also used for two-legged animals.

Stotting (applies to the Mule Deer only): Describes the action of taking off from the ground and landing on all four feet at once, in pogo-stick fashion.

Trotting: Faster than walking, slower than running. The diagonally opposite limbs move simultaneously; that is, the right forefoot with the left hind, then the left forefoot with the right hind. This gait is the natural one for canids (dog family), short-tailed shrews and voles.

hind print   fore print

Canids may use **side-trotting**, a fast trotting in which the hind end of the animal shifts to one side. The resulting track pattern has paired tracks, with all the fore prints on one side and all the hind prints on the other.

Walking: A slow gait in which each foot moves independently of the others, resulting in an alternating track pattern. This gait is common for felines (cat family) and deer, as well as wide-bodied animals, such as bears and porcupines. The term is also used for two-legged animals.

Bobcat

Other Tracking Terms:

Dewclaws: Two small, toe-like structures set above and be-hind the main foot of most hoofed animals.

Direct Register: The hind print falls directly on the fore print.

Double Register: The hind print overlaps the fore print only slightly or falls beside it, so that both can be seen at least in part.

Dragline: A line left in snow or mud by a foot or the tail dragging over the surface.

Gallop Group: A track pattern of four prints made at a gal-lop, usually with hind feet registering in front of forefeet.

Height: Taken at the animal's shoulder.

Length: The animal's body length from head to rump, not including the tail, unless otherwise indicated.

Metacarpal Pad: A small pad near the palm pad or between the palm pad and heel on the forefeet of bears and members of the weasel family.

Print (also called '***track***'): Fore and hind prints are treated individually. Print dimensions given are 'length' (including claws—maximum values may represent occasional heel register for some animals) and 'width.' A group of prints made by each of the animal's feet makes up a track pattern.

Register: To leave a mark—said about a foot, claw or other part of an animal's body.

Retractable: Describes claws that can be pulled in to keep them sharp, as with the cat family; these claws do not register in the prints. Foxes have semi-retractable claws.

Sitzmark: The mark left on the ground by an animal falling or jumping from a tree.

Straddle: The total width of the trail, all prints considered.

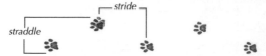

Stride: For consistency among different animals, the stride is taken as the distance from the centre of one print (or print group) to the centre of the next one. Some books may use the term 'pace.'

Track: Same as '***print***.'

Track Pattern: The pattern left after each foot registers once; a set of prints, such as a gallop group.

Trail: A series of track patterns; think of it as the path of the animal.

MAMMALS

River Otter

Bison

Fore and Hind Prints
Length: 10–15 cm (4–6 in)
Width: 10–15 cm (4–6 in)
Straddle
25–55 cm (10–21 in)
Stride
Walking: 35–80 cm (14–32 in)
Size (bull>cow)
Height: 1.5–1.8 m (5–6 ft)
Length: 3–3.7 m (10–12 ft)
Weight
Male: 360–900 kg (800–2000 lb)
Female: 320–500 kg (700–1100 lb)

walking

BISON (Buffalo)
Bos bison

Bison, frequently called 'buffalo,' once roamed North America in numbers estimated at 70 million. As few as 1500 remained after the wholesale slaughter of the Bison during the nineteenth century, and thereafter a major effort began to save this magnificent beast from extinction. Today, as many as 100,000 Bison roam selected protected areas and public and private ranches, resulting in a very scattered distribution. Do not be fooled by a Bison's calm exterior: a hefty bull Bison can inflict serious injury!

In a Bison's alternating walking pattern, the slightly smaller hind foot usually registers on or near the fore. On firm ground, only the outer edge of the hoof may register, but in soft mud or snow the whole foot registers—perhaps the dewclaws too—and foot drag is common. The abundant 'pies' may be mistaken for the dung of domestic cattle. Additional signs of Bison include rubbing posts, tufts of distinctive brown hair hanging from trees and the large pits in which Bison wallow.

Similar Species: Domestic Cattle (*Bos* spp.) prints are similar. On firm surfaces, Horse (p. 32) prints can resemble Bison prints.

17

Caribou

Fore and Hind Prints
Length: 7.5–12 cm (3–4.7 in)
Length with dewclaws:
    to 20 cm (8 in)
Width: 10–15 cm (4–6 in)

Straddle
23–35 cm (9–14 in)

Stride
Walking: 40–80 cm (16–32 in)
Running: to 1.5 m (5 ft)
Group length: to 2.7 m (9 ft)

Size (buck>doe)
Height: 1.1–1.2 m (3.5–4 ft)
Length: 2–2.6 m (6.5–8.5 ft)

Weight
70–270 kg (150–600 lb)

*walking
(in snow)*

*walking
(on hard ground)*

CARIBOU
Rangifer tarandus

There are only a few places in British Columbia where you might be privileged to see the elegant antlers of the male Caribou, a true wilderness animal. The Caribou's sensitivity to human encroachment has reduced its range to mountain parks, where it prefers to feed in groups above the tree line.

In winter, the soft inner sole of a Caribou's hoof hardens and shrinks, leaving a firm outer wall that makes neat circles on firmer surfaces. The hooves spread wide to act like snowshoes, leaving distinctive large, rounded prints. Big dewclaws, which help caribou to distribute their weight on snow, register behind the forefeet, but rarely the hind feet. The faster a Caribou runs, the more perpendicular its dewclaws are to the direction of travel. In snow, foot drag is common; examine it to see how Caribou swing their legs as they walk. Look for scrape marks where Caribou have dug for lichens hidden underneath the snow.

Similar Species: By virtue of the Caribou's print shape and size and its range, there is little chance for confusion.

 19

Moose

Fore and Hind Prints
Length: 10–18 cm (4–7 in)
Length with dewclaws: to 28 cm (11 in)
Width: 9–15 cm (3.5–6 in)

Straddle
22–50 cm (8.5–20 in)

Stride
Walking: 45–90 cm (1.5–3 ft)
Trotting: to 1.2 m (4 ft)

Size (bull>cow)
Height: 1.5–2 m (5–6.5 ft)
Length: 2.1–2.6 m (7–8.5 ft)

Weight
270–500 kg (600–1100 lb)

walking

MOOSE
Alces alces

The impressive male Moose, largest of the deer, has a massive rack of antlers. Moose are usually solitary, though you may see a cow with her calf. Despite their placid appearance, both the bull and the cow will occasionally charge humans if approached.

The ungainly shaped Moose moves gracefully, leaving a neat, alternating walking pattern. The hind feet directly register or double register on the fore prints. Long legs allow for easy movement in snow; where the print is deeper than 3 cm (1.2 in), the dewclaws—which give extra support for the animal's great weight—do register, but far back from the main print. In summer, look for tracks in mud beside ponds and other wet areas, where Moose especially like to feed; they are excellent swimmers. In winter Moose feed in willow flats and coniferous forests, leaving a distinct browseline (highline). Ripped stems and scraped bark, 1.8 m (6 ft) or more above the ground, are additional signs that Moose have been around.

Similar Species: Elk (p. 22) prints are rounder and smaller, with a narrower straddle and more foot drag, but may be easily confused with juvenile Moose prints.

Fore and Hind Prints
Length: 8–13 cm (3.2–5 in)
Width: 6.5–11 cm (2.5–4.5 in)
Straddle
18–30 cm (7–12 in)
Stride
Walking: 40–85 cm (16–34 in)
Galloping: 1–2.4 m (3.3–8 ft)
Group length: to 1.9 m (6.3 ft)
Size (bull>cow)
Height: 1.2–1.5 m (4–5 ft)
Length: 2–3 m (6.5–10 ft)
Weight
225–450 kg (500–1000 lb)

gallop print   *walking*

ELK (Wapiti)
Cervus elaphus

Elk are common in meadows and open forests of the mountains and foothills. Female Elk and young are often seen in social herds. They like to feed in forest openings and lush meadows. Stags, who prefer to go solo, are easily recognized by their magnificent racks of antlers and distinctive bugling in late August. Herds move into valleys when winter sets in. A good place to look for Elk tracks is in the soft mud beside summer ponds, where Elk like to drink and sometimes splash around.

Elk leave a neat, alternating walking pattern of large, rounded prints, often in well-worn winter paths. The hind foot will sometimes double register slightly in front of the fore print. In deeper snow, or if an Elk gallops (with its toes spread wide), the dewclaws may register.

Similar Species: Smaller deer (pp. 24–27) leave similar track patterns, but with generally shorter, narrower prints.

Mule Deer

Fore and Hind Prints
Length: 5–8.5 cm (2–3.3 in)
Width: 4–6.5 cm (1.6–2.5 in)

Straddle
13–25 cm (5–10 in)

Stride
Walking: 25–60 cm (10–24 in)
Stotting: 2.7–5.8 m (9–19 ft)

Size (buck>doe)
Height: 90–110 cm (3–3.5 ft)
Length: 1.2–2 m (4–6.5 ft)

Weight
45–200 kg (100–450 lb)

walking   *stot group*

MULE DEER
(Black-tailed Deer)
Odocoileus hemionus

The widespread Mule Deer is frequently seen in meadows and open woodlands and on desert plains. In winter it moves down from higher terrain to warmer south-facing slopes and sagebrush flats, where it can feed without having to contend with deep snow. It prefers to stay in small groups and frequently uses the same well-worn path in winter.

The Mule Deer has a neat alternating walking pattern, with the hind foot registering on the fore print. Mule Deer prints are heart-shaped and sharply pointed. In deeper snow or when a deer is moving quickly, its prints show dewclaws; on the fore print they are closer to the toes.

At high speed this deer has a unique gait—stotting—in which it jumps with all its feet leaving and striking the ground at once. Stotting track patterns show how the toes spread to distribute the weight and give better footing.

Similar Species: White-tailed Deer (p. 26) prefer denser cover and have a different gallop pattern with a shorter stride. Elk (p. 22) have longer, wider prints.

White-tailed Deer

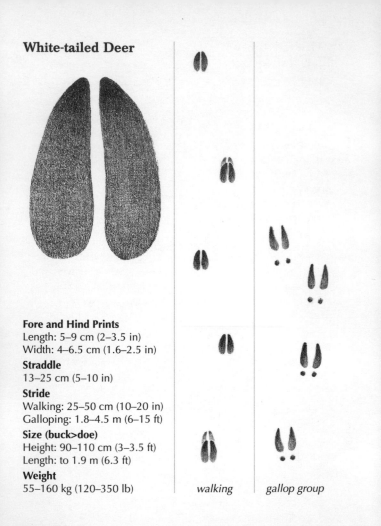

Fore and Hind Prints
Length: 5–9 cm (2–3.5 in)
Width: 4–6.5 cm (1.6–2.5 in)

Straddle
13–25 cm (5–10 in)

Stride
Walking: 25–50 cm (10–20 in)
Galloping: 1.8–4.5 m (6–15 ft)

Size (buck>doe)
Height: 90–110 cm (3–3.5 ft)
Length: to 1.9 m (6.3 ft)

Weight
55–160 kg (120–350 lb)

walking   *gallop group*

WHITE-TAILED DEER
Odocoileus virginianus

The keen hearing of this deer guarantees that it knows about you before you know about it. Frequently, all that we see is its conspicuous white tail in the distance as it gallops away, earning this deer the nickname 'flagtail.' This adaptable deer may be found in small groups at the edges of forests and in brushlands, throughout much of the province. White-tailed Deer can be common around ranches and residential areas.

Their prints are heart-shaped and pointed. They walk in an alternating track pattern, with the hind feet direct or double registering on the fore prints. In snow or when a deer gallops on soft surfaces, the dewclaws register. This flighty deer gallops in the usual style, hind prints located in front of fore prints, with toes spread wide for better footing.

Similar Species: Mule Deer (p. 24) make similar prints but prefer more open terrain and stot (not gallop) when moving at high speed. Elk (p. 22) prints are longer and wider.

Mountain Goat

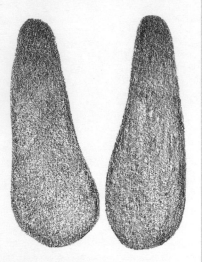

Fore and Hind Prints
Length: 6.5–9 cm (2.5–3.5 in)
Width: 5–8.5 cm (2–3.3 in)
Straddle
17–30 cm (6.5–12 in)
Stride
Walking: 25–48 cm (10–19 in)
Size (billy>nanny)
Height: 90–110 cm (3–3.5 ft)
Length: 1.5–1.8 m (5–6 ft)
Weight
45–140 kg (100–300 lb)

walking

MOUNTAIN GOAT
Oreamnos americanus

Spotting the dazzling white coat of a Mountain Goat is truly a high-mountain wilderness experience. This goat has a preference for high terrain, usually rocky slopes above the tree line. The Mountain Goat's keen vision and its remote setting make it difficult to approach; its tracks in snow may be your best clue that it is around.

The goat's tracks show that it has long, widely spreading toes that produce a squarish print. The hard rim and soft middle of the foot help this agile goat clamber over the most unlikely crags at remarkable speeds, but even the Mountain Goat can make a fatal mistake! In deeper snow, the feet may leave draglines and the dewclaws may register. The alternating walking pattern shows a double register, hind print on top of fore.

Similar Species: Mule Deer (p. 24), White-tailed Deer (p. 26) and Bighorn Sheep (p. 30) prints are often narrower and more pointed. Few deer or sheep climb as high as the Mountain Goat but, in severe weather, this goat may come down into deer territory.

 29

Bighorn Sheep

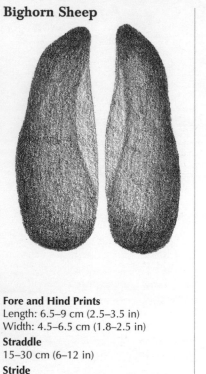

Fore and Hind Prints
Length: 6.5–9 cm (2.5–3.5 in)
Width: 4.5–6.5 cm (1.8–2.5 in)
Straddle
15–30 cm (6–12 in)
Stride
Walking: 35–60 cm (14–24 in)
Size (ram>ewe)
Height: 75–110 cm (2.5–3.5 ft)
Length: 1.2–2 m (4–6.5 ft)
Weight
34–120 kg (75–270 lb)

walking

BIGHORN SHEEP
(Mountain Sheep)
Ovis canadensis

In late fall, the loud crack of two majestic rams head-butting one another can be heard for a great distance. To watch the rut is an awe-inspiring experience, but a rare one. The Mountain Sheep prefers high, open meadows and scree slopes in the mountains, moving into valleys only in winter. It tends to avoid forested areas, and is not quite as bold as the Mountain Goat (p. 28) on craggy cliffs.

The squarish print is pointed toward the front. The outer edge of the hoof is hard and the inner part is soft, giving a good grip on tricky terrain. The neat alternating walking pattern is a direct or double register of the hind foot on top of the fore print. When this sheep runs, its toes spread wide. Finding several trails together is likely, because this sheep likes to travel in herds. The trails may lead you to sheep beds—hollows dug into the snow that are used many times, often with a large accumulation of droppings.

Similar Species: Dall's Sheep (*Ovis dalli*) is indistinguishable by trail, but it inhabits more northerly parts of the province. Domestic Sheep (*Ovis* spp.) prints are similar. Mountain Goats have prints that are wider at the toe and rarely run. Deer (pp. 24–27) prints are more heart-shaped.

Horse

Fore Print
(hind print is slightly smaller)
Length: 11–15 cm (4.5–6 in)
Width: 11–14 cm (4.5–5.5 in)
Stride
Walking: 43–70 cm (17–27 in)
Size
Height: to 1.8 m (6 ft)
Weight
to 680 kg (1500 lb)

walking

HORSE
Equus caballus

This popular animal has unmistakable prints; it deserves mention because back-country use of the Horse means that you can expect its tracks to show up almost anywhere.

Unlike any other animal discussed in this book, the Horse has only one huge toe, which leaves an oval print. A distinctive feature is the 'frog,' or V-shaped mark, at the base of the print. When the Horse is shod, the horseshoe shows up clearly as a firm wall at the outside of the print. Not all horses will be shod, so do not expect to see this outer wall on every horse print. A typical, unhurried horse trail is an alternating walking pattern, with the hind feet registering on or behind the slightly larger fore prints. Horses are capable of a range of speeds—up to a full gallop—but most recreational horseback riders take a more leisurely outlook on life, preferring to walk their horses and soak up the mountain views!

Similar Species: Mules (rarely shod) have smaller prints.

Grizzly Bear

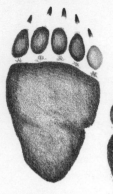

hind

fore

Fore Print
Length: 13–18 cm (5–7 in)
Width: 10–15 cm (4–6 in)

Hind Print
Length: 23–30 cm (9–12 in)
Width: 13–18 cm (5–7 in)

Straddle
25–50 cm (10–20 in)

Stride
Walking: 60–100 cm (24–40 in)

Size (male>female)
Height: 90–110 cm (3–3.5 ft)
Length: 1.8–2.1 m (6–7 ft)

Weight
90–390 kg (200–850 lb)

*walking
(fast)*

GRIZZLY BEAR
Ursus arctos

The infrequently seen but magnificent and imposing Grizzly Bear symbolizes mountain wilderness for many of us. It is very sparsely distributed throughout the mountains of British Colombia. Grizzly Bears prefer open country and valley bottoms, and they are sensitive to human activity. In winter, bears enter a deep slumber, so few of their tracks are seen.

Each of the Grizzly's huge prints will show four or five toes, with very long claws and a small heel pad on the fore print. A solid rear heel makes for a sturdy hind print. The toes are closely set in a line, with the inner toe the smallest. The usual walking track pattern shows the hind foot registering in front of the forefoot, although a slower gait results in a trail like the Black Bear's (p. 36). The Grizzly occasionally gallops. Well-worn trails may lead to digs, trees with claw marks high up on the trunks or even the cache of a carcass. Take care if you find a cache—the unpredictable Grizzly is likely to be nearby.

Similar Species: A Black Bear's prints are smaller, with shorter claw marks and toes arranged more in an arc, and its territorial tree-scratchings are lower on the trunk.

Black Bear

hind

fore

Fore Print
Length without heel:
    10–16 cm (4–6.3 in)
Width: 9.5–14 cm (3.8–5.5 in)
Hind Print
Length: 15–18 cm (6–7 in)
Width: 9–14 cm (3.5–5.5 in)
Straddle
23–38 cm (9–15 in)
Stride
Walking: 43–58 cm (17–23 in)
Size (male>female)
Height: 90–110 cm (3–3.5 ft)
Length: 1.5–1.8 m (5–6 ft)
Weight
90–270 kg (200–600 lb)

walking
(slow)

BLACK BEAR
Ursus americanus

The Black Bear is widespread in forested areas throughout British Columbia but, because it sleeps deeply in winter, do not expect to encounter its tracks during the colder months. Finding fresh bear tracks can be a thrill, but take care, as the bear may be just around the corner. Never underestimate the potential power of a surprised bear!

Black Bear prints somewhat resemble smaller human prints, but are wider and show claw marks. The small inner toe rarely registers. The forefoot's small heel pad often shows, and the hind foot has a big heel. The bear's slow walk results in a slightly pigeon-toed double register of the hind foot over the fore print. More frequently, at a faster pace, the hind oversteps the fore, as shown for the Grizzly Bear (p. 34). When a bear runs, the two hind feet register in front of the forefeet in an extended cluster. Along well-worn bear paths, look for 'digs'—patches of dug-up earth—and 'bear trees' whose scratched bark shows that Black Bears climb.

Similar Species: Grizzly prints (often larger) show toes set more in a line and closer together as well as longer claws.

Grey Wolf

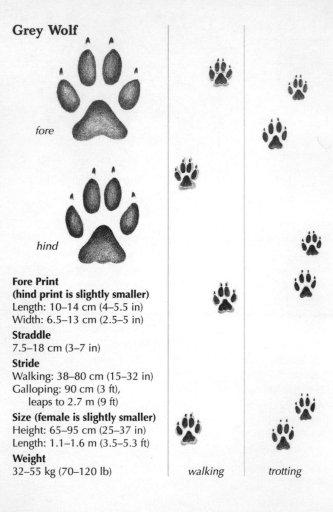

fore

hind

Fore Print
(hind print is slightly smaller)
Length: 10–14 cm (4–5.5 in)
Width: 6.5–13 cm (2.5–5 in)

Straddle
7.5–18 cm (3–7 in)

Stride
Walking: 38–80 cm (15–32 in)
Galloping: 90 cm (3 ft),
    leaps to 2.7 m (9 ft)

Size (female is slightly smaller)
Height: 65–95 cm (25–37 in)
Length: 1.1–1.6 m (3.5–5.3 ft)

Weight
32–55 kg (70–120 lb)

walking

trotting

GREY WOLF
(Timber Wolf)
Canis lupus

The soulful howl of the wolf epitomizes the outdoor experience, but few people ever hear it. Your best bet for hearing a wolf is in national parks or remote, undisturbed areas. The largest of the wild dogs, the Grey Wolf may travel in packs or alone, but is rarely seen.

A wolf leaves a straight, direct-registering alternating track pattern, with the smaller hind print on top of the fore print. It has large, oval prints that each show all four claws. The lobing on the fore print heel pads is different from that of the hind ones. A wolf trail in the snow is likely the trail of several wolves, because in deep snow wolves sensibly follow their leader, sometimes dragging their feet. When a wolf trots, notice how the hind print has a slight lead and falls to one side, giving an unbalanced appearance. Wolves and Coyotes (p. 40) gallop in the same way.

Similar Species: Domestic Dog (*Canis familiaris*) prints, rarely as large as a wolf's, fall in a haphazard track pattern with a less direct register, and the inner toes tend to spread out more. A Wolverine (p. 62) print in which the inner toe does not register may be confused with a wolf's, but the pad shapes are very different.

Coyote

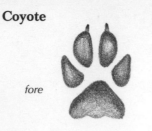

fore

hind

Fore Print
(hind print is slightly smaller)
Length: 6–8 cm (2.4–3.2 in)
Width: 4–6 cm (1.6–2.4 in)

Straddle
Walking: 10–18 cm (4–7 in)

Stride
Walking: 30–40 cm (12–16 in)
Trotting: 48–60 cm (19–24 in)
Galloping: 0.8–3 m (2.5–10 ft)

Size (female is slightly smaller)
Height: 58–65 cm (23–26 in)
Length: 80–100 cm (32–40 in)

Weight
9–23 kg (20–50 lb)

walking

gallop group

COYOTE (Brush Wolf, Prairie Wolf)
Canis latrans

This widespread and adaptable canine prefers open grasslands and woodlands. It hunts rodents and larger prey, either on its own, with a mate or in a family pack. A Coyote also occasionally develops an interesting cooperative relationship with a Badger (p. 74), so you might find their tracks together where they have been digging for ground squirrels.

The oval fore prints are slightly larger than the hind prints. Note the difference between the fore heel pad and the hind heel pad, which rarely registers clearly. Claw marks are usually absent for the two outer toes. A Coyote's tail hangs down, leaving a dragline in deep snow. Coyotes typically walk or trot in an alternating pattern—the walk having a wider straddle—often with the hind feet registering in front of the forefeet. The trotting trail made by a Coyote is often very straight. When a Coyote gallops, the hind feet fall in front of the forefeet; the faster it goes, the straighter the gallop group.

Similar Species: A Domestic Dog's (*Canis familiaris*) less-oval prints splay more, and its trail is erratic. Foot hairs make Red Fox (p. 42) prints (usually smaller) less clear.

Red Fox

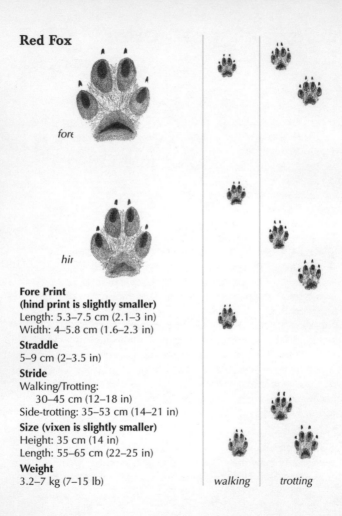

fore

hir

Fore Print
(hind print is slightly smaller)
Length: 5.3–7.5 cm (2.1–3 in)
Width: 4–5.8 cm (1.6–2.3 in)

Straddle
5–9 cm (2–3.5 in)

Stride
Walking/Trotting:
    30–45 cm (12–18 in)
Side-trotting: 35–53 cm (14–21 in)

Size (vixen is slightly smaller)
Height: 35 cm (14 in)
Length: 55–65 cm (22–25 in)

Weight
3.2–7 kg (7–15 lb)

walking   *trotting*

RED FOX
Vulpes vulpes

   This beautiful
and notoriously
cunning fox is found throughout British Columbia, but prefers mountainous forests and open areas. A very adaptable
and intelligent animal, it is secretive and largely nocturnal.

   The Red Fox has very hairy feet, so the finer details of
its prints are obscured—only parts of the toes and heel pads
show. A significant feature on the fore prints of this fox is the
horizontal or slightly curved bar across the heel pad.

   A trotting Red Fox leaves a distinctive straight trail of
alternating prints, with the hind foot direct registering on
the wider fore print. When a fox side-trots, the paired prints
show the hind print falling to one side of the fore print in
typical canid fashion. Foxes gallop like Coyotes (p. 40). The
faster the gallop, the straighter the group.

Similar Species: Domestic Dog (*Canis familiaris*) prints are
of similar size but lack the bar on the heel pad, and show
a shorter stride with a less direct trail. Small Coyote prints
are similar, but with a wider straddle and more bulbous toe
registrations.

Mountain Lion

fore

hind

Fore Print
(hind print is slightly smaller)
Length: 7.5–11 cm (3–4.5 in)
Width: 8.5–12 cm (3.3–4.8 in)

Straddle
20–30 cm (8–12 in)

Stride
Walking: 33–80 cm (13–32 in)
Bounding: to 3.7 m (12 ft)

Size
Height: 65–80 cm (26–32 in)
Length: 1.1–1.5 m (3.5–5 ft)

Weight
32–90 kg (70–200 lb)

walking (fast)

MOUNTAIN LION
(Puma, Cougar)
Felis concolor

The Mountain Lion is shy, elusive and nocturnal in nature, so finding its tracks is usually the best that trackers can hope for. Mountain Lions are spread widely but sparsely because of their need for a big home territory.

Mountain Lion prints tend to be wider than long and the retractable claws never register. In winter, thick fur makes the prints look much larger, and it may obscure the two lobes on the front of the heel pad. When a Mountain Lion walks, the hind foot either directly or double registers on the larger fore print. As its walking pace increases, the hind print tends to fall in front of the fore print. In snow, the long, thick tail may leave a dragline that can obscure some of the print detail. A Mountain Lion seldom gallops but, when it needs to catch prey, it is capable of long bounds.

Similar Species: Bobcat (p. 48) prints may be confused with juvenile Mountain Lion prints but are usually smaller, clearer and sunk into the snow more. Lynx (p. 46) prints are similar in size, but Lynx have a narrower straddle, a shorter stride and no tail drag, and they do not sink as deep in snow.

 45

Lynx

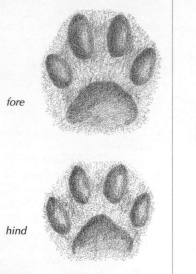

fore

hind

Fore Print
(hind print is slightly smaller)
Length: 9–11 cm (3.5–4.5 in)
Width: 9–12 cm (3.5–4.8 in)
Straddle
15–23 cm (6–9 in)
Stride
Walking: 30–70 cm (12–28 in)
Size
Length: 75–90 cm (2.5–3 ft)
Weight
7–14 kg (15–30 lb)

walking

LYNX
Lynx canadensis

This large cat
would be a thrill to
see, but it usually
eludes humans by
living in dense
forests. The
Lynx is sensi-
tive to human
interference, so it is abundant only in remote and undisturbed parts of the province. With its huge feet and rela-
tively lightweight body, the Lynx stays on top of the snow
as it pursues its main prey, the Snowshoe Hare (p. 78).

This cautious walker leaves an alternating track pattern
with neat, direct registers of the hind foot on top of the fore
print. Thick fur on the feet often results in prints that are
big, round depressions with no detail. In deeper snow, the
print may be extended by 'handles' off to the rear. However
deep the snow, this cat sinks no more than 20 cm (8 in) and
it rarely drags its feet. A Lynx is more likely to bound than
to run. Its curious nature results in a meandering trail that
may lead to a partially buried food cache.

Similar Species: Mountain Lions (p. 44) have similar-
sized, clearer prints, sink deeper in snow and have a wider
straddle. Bobcat (p. 48) prints are smaller and the trail may
show draglines. Coyote (p. 40), Domestic Dog (*Canis familiaris*) and Red Fox (p. 42) prints are narrower than they are
long and show claw marks.

Bobcat

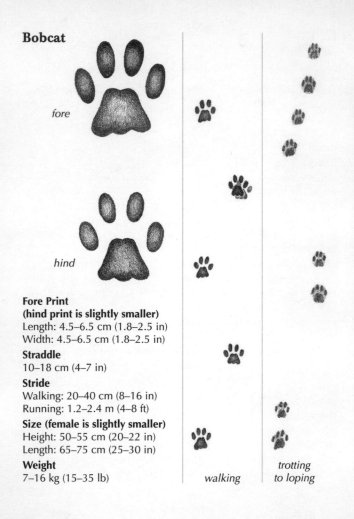

fore

hind

Fore Print
(hind print is slightly smaller)
Length: 4.5–6.5 cm (1.8–2.5 in)
Width: 4.5–6.5 cm (1.8–2.5 in)

Straddle
10–18 cm (4–7 in)

Stride
Walking: 20–40 cm (8–16 in)
Running: 1.2–2.4 m (4–8 ft)

Size (female is slightly smaller)
Height: 50–55 cm (20–22 in)
Length: 65–75 cm (25–30 in)

Weight
7–16 kg (15–35 lb)

walking

*trotting
to loping*

BOBCAT (Wildcat)
Lynx rufus

The seldom-seen Bobcat, widely distributed in British Columbia, is a stealthy hunter that usually pursues its prey in the secret of the night. The Bobcat is very adaptable; it can leave tracks anywhere from wild mountainsides to chaparral and even into residential areas.

The hind foot usually registers directly on the larger fore print in the walking pattern. The fore prints especially show the asymmetrical shape of the track. The front part of the heel pad has two lobes and the rear part three. As a Bobcat picks up speed, its trail becomes an ambling pattern of paired prints, the hind leading the fore. At even greater speeds, it leaves four-print groups in a lope pattern. A Bobcat's feet leave draglines in deep snow. Unlike the trails of wild dogs, the Bobcat's trail meanders. Half buried scat along the trail is a sign of a Bobcat marking its territory.

Similar Species: Juvenile Mountain Lion (p. 44) and Lynx (p. 46) prints can be similar. A large Domestic Cat (p. 50) has similar prints, but a shorter stride and a narrower straddle. Fisher (p. 64) hind prints that lack the fifth toe may seem similar—look for mustelid habits. Coyote (p. 40), Domestic Dog (*Canis familiaris*) and Red Fox (p. 42) prints are narrower than long and show claw marks; also, the fronts of their footpads are once-lobed.

Domestic Cat

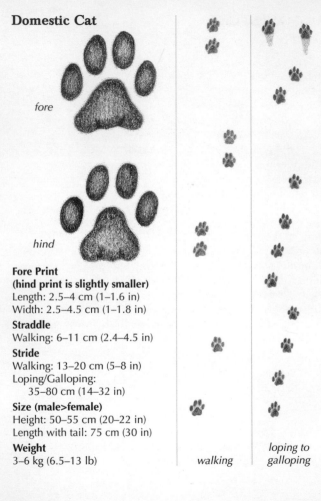

fore

hind

Fore Print
(hind print is slightly smaller)
Length: 2.5–4 cm (1–1.6 in)
Width: 2.5–4.5 cm (1–1.8 in)

Straddle
Walking: 6–11 cm (2.4–4.5 in)

Stride
Walking: 13–20 cm (5–8 in)
Loping/Galloping:
    35–80 cm (14–32 in)

Size (male>female)
Height: 50–55 cm (20–22 in)
Length with tail: 75 cm (30 in)

Weight
3–6 kg (6.5–13 lb)

walking

loping to galloping

DOMESTIC CAT
(House Cat)
Felis catus

The familiar Domestic Cat is so abundant in residential areas that its tracks can show up almost any place where there are people. From time to time, cats are abandoned and roam further afield. They are known as 'feral cats,' because they lead a pretty wild and independent existence. Domestic Cats can come in many shapes, sizes and colours.

As with all members of the cat family, a Domestic Cat's fore and hind prints both show four toe pads. Its retractable claws do not register—they are kept clean and sharp for catching prey. Cat prints usually show a slight asymmetry, with one toe leading the others. The hind print is slightly smaller than the fore print. A Domestic Cat makes a neat, alternating walking trail, usually in direct register, as one would expect from this animal's fastidious nature. When a cat picks up speed, it leaves clusters of four prints, the hind feet registering in front of the forefeet.

Similar Species: A small Bobcat (p. 48) may leave tracks similar to a very large Domestic Cat's. Red Fox (p. 42) and Domestic Dog (*Canis familiaris*) prints show claw marks.

51

Raccoon

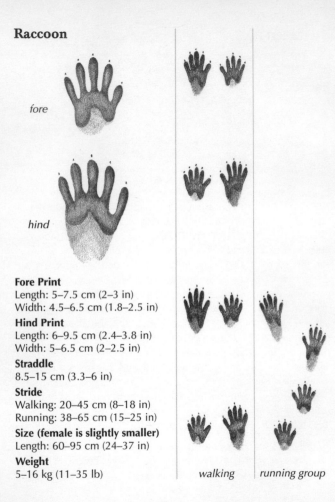

fore

hind

Fore Print
Length: 5–7.5 cm (2–3 in)
Width: 4.5–6.5 cm (1.8–2.5 in)
Hind Print
Length: 6–9.5 cm (2.4–3.8 in)
Width: 5–6.5 cm (2–2.5 in)
Straddle
8.5–15 cm (3.3–6 in)
Stride
Walking: 20–45 cm (8–18 in)
Running: 38–65 cm (15–25 in)
Size (female is slightly smaller)
Length: 60–95 cm (24–37 in)
Weight
5–16 kg (11–35 lb)

walking   *running group*

RACCOON
Procyon lotor

The inquisitive Raccoon
is adored for its distinctive face mask, yet disliked for its
boundless curiosity—often demonstrated in residential
garbage cans. Raccoons are common in southern parts of
British Columbia, and a good place to look for their tracks
is near water at low elevations. Raccoons like to rest in trees
and they usually den up for the colder months.

The Raccoon's unusual print, showing five well-formed
toes, looks like a human handprint. The small claws appear
as dots. Its highly dexterous forefeet rarely leave heel prints,
but its hind prints, which are generally much clearer, do
show heels. The Raccoon's peculiar walking track
pattern shows a left fore print next to (or just in front of)
a right hind print and vice versa. On the rare occasions
when a Raccoon is out in deep snow, it may use a direct-
registering walk. Raccoons occasionally run, leaving clusters
where the two hind prints fall in front of the fore prints.

Similar Species: Opossum (p. 54) prints that are unclear
may look similar, but the Opossum drags its tail. Direct-
registering Raccoon tracks in snow may be confused with
a walking Fisher's (p. 64) or River Otter's (p. 60), but these
mustelids do not walk very far or often.

53

Opossum

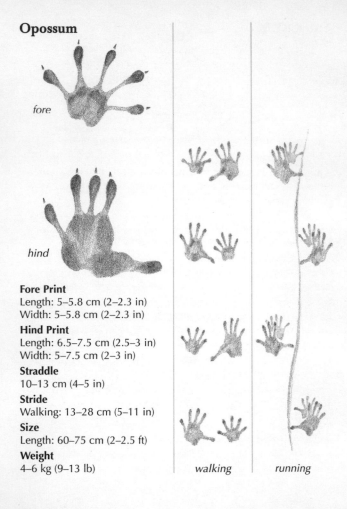

fore

hind

Fore Print
Length: 5–5.8 cm (2–2.3 in)
Width: 5–5.8 cm (2–2.3 in)
Hind Print
Length: 6.5–7.5 cm (2.5–3 in)
Width: 5–7.5 cm (2–3 in)
Straddle
10–13 cm (4–5 in)
Stride
Walking: 13–28 cm (5–11 in)
Size
Length: 60–75 cm (2–2.5 ft)
Weight
4–6 kg (9–13 lb)

walking

running

OPOSSUM
Didelphis virginiana

This slow-moving, nocturnal
marsupial is found in the southwestern
coastal region of British Columbia. Though it
occupies many types of habitats, it prefers open
woodland or brushland around waterbodies. It is
quite tolerant of residential areas. Opossum tracks
can often be seen in mud near the water and in snow
during the warmer months of winter—Opossums
tend to den up during severe weather.

Opossums are excellent climbers, so do not be
surprised if their tracks lead to a tree. They have two
walking habits: the common alternating pattern, with the
hind feet registering on the fore prints, and a Raccoon-like
paired-print pattern, with the hind print next to the oppos-
ing fore print. The very distinctive, long, inward-pointing
thumb of the hind foot does not make a claw mark.
In snow, the dragline of the long, naked tail may be stained
with blood; unfortunately, this thinly haired animal is not
well adapted to the cold and frequently suffers frostbite.

Similar Species: Prints in which the distinctive thumbs do
not show, as in sand or fine snow, may be mistaken for a
Raccoon's (p. 52).

Harbour Seal

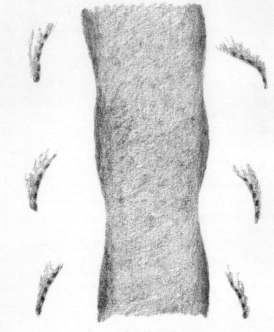

beach track

Size (male>female)
Length: 1.2–1.8 m (4–6 ft)
Weight
80–140 kg (180–300 lb)

HARBOUR SEAL
Phoca vitulina

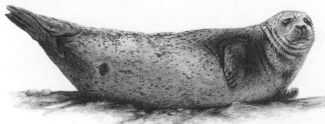

This common seal frequents the isolated beaches of the coast. One of the smaller seals, it is quite shy, so it will usually slide off its rocky sentry post into the sea to escape curious naturalists. Look in sandy and muddy areas between rocky platforms for its tracks—their location and large size make them unmistakable. Sometimes a Harbour Seal will work its way up a river, so you may also find its tracks along riverbanks.

Seals are not the most graceful of animals on land, but they are unrivalled in the water. Their heavy, fat bodies and flipper-like feet can leave messy tracks: a wide trough (made by their cumbersome bellies) with dents alongside (made as the seals pushed themselves along with their forefeet). Look for the marks made by the large nails.

Similar Species: The Northern Fur Seal (*Callorhinus ursinus*), which can weigh twice as much, also frequents the mainland coast. The Northern Sea Lion (*Eumetopias jubatus*) is sometimes seen along the shore, too; it can weigh seven times as much as the little Harbour Seal. Other seals and sea lions can be found on the remote offshore islands.

Sea Otter

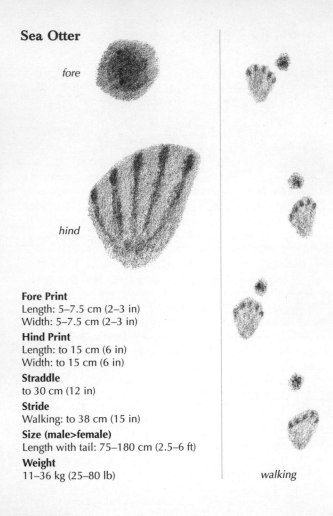

fore

hind

Fore Print
Length: 5–7.5 cm (2–3 in)
Width: 5–7.5 cm (2–3 in)
Hind Print
Length: to 15 cm (6 in)
Width: to 15 cm (6 in)
Straddle
to 30 cm (12 in)
Stride
Walking: to 38 cm (15 in)
Size (male>female)
Length with tail: 75–180 cm (2.5–6 ft)
Weight
11–36 kg (25–80 lb)

walking

SEA OTTER
Enhydra lutris

As fun-loving as its freshwater cousin, the River Otter (p. 60), this large inhabitant of coastal waters is so suited to the sea that it seldom comes onto land. If you do catch sight of a Sea Otter on land, it might well be cavorting with sea lions, or waiting out a violent storm. If you are lucky, you may hear the tapping sound of an otter knocking on a shellfish with a stone to break it open. Sea Otter scat is recognizable by its crumbly texture and fragments of shell.

Although the Sea Otter rarely comes ashore, its tracks are most likely on beaches where kelp grows offshore. Like the seals, this otter is not as graceful on land as it is in the sea, because its webbed hind feet are almost flippers. The roundish fore print is much smaller than the hind print, and the trail usually shows an alternating walking pattern.

Similar Species: The Harbour Seal (p. 56) also has flipper-like feet, but it leaves an obvious trough with its belly and it has much larger forefeet.

 59

River Otter

fore

hind

Fore Print
Length: 6.5–9 cm (2.5–3.5 in)
Width: 5–7.5 cm (2–3 in)

Hind Print
Length: 7.5–10 cm (3–4 in)
Width: 5.8–8.5 cm (2.3–3.3 in)

Straddle
10–23 cm (4–9 in)

Stride
Loping: 30–70 cm (12–27 in)

Size
(female is two-thirds the size of male)
Length with tail: 90–130 cm (3–4.3 ft)

Weight
4.5–11 kg (10–25 lb)

loping (fast)

RIVER OTTER
Lontra canadensis

No animal knows how to have more fun than a River Otter. If you are lucky enough to watch one at play, you will not soon forget the experience. Well-adapted for the aquatic environment, this otter is widespread, living along streams and other waterbodies. Within an otter's home territory you are sure to find a wealth of evidence along the riverbanks. Otters are occasionally found in a forest, but they are usually going directly to another waterbody.

In soft mud, the River Otter's five-toed feet register evidence of webbing, the hind feet more than the forefeet. The inner toes are set slightly apart. The forefoot's metacarpal pad may register, lengthening the print. Otter trails are very variable: they show the typical 2x2 loping of the mustelids and, with faster gaits, show groups of four and three prints. The thick, heavy tail often leaves a mark over the prints. Otters love to slide in snow, often down riverbanks, leaving troughs nearly 30 cm (12 in) wide. In summer they roll and slide on grass and mud.

Similar Species: Mink (p. 68) prints are similar but less than half the size; The Fisher (p. 64) is found in forested regions, usually has hairy feet with the forefeet larger than the hind feet and does not show conspicuous tail drag.

61

Wolverine

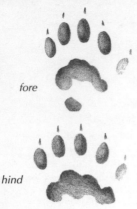

fore

hind

Fore Print
Length with heel: 10–19 cm (4–7.5 in)
Width: 10–13 cm (4–5 in)

Hind Print
Length: 9–10 cm (3.5–4 in)
Width: 10–13 cm (4–5 in)

Straddle
18–23 cm (7–9 in)

Stride
Walking: 7.5–30 cm (3–12 in)
Loping: 25–100 cm (10–40 in)

Size (female is slightly smaller)
Height: 40 cm (16 in)
Length: 80–120 cm (32–46 in)

Weight
8–21 kg (18–45 lb)

loping (slow)

WOLVERINE
Gulo gulo

The reputation of the robust and powerful Wolverine, the largest of the mustelids, has earned it many nicknames, such as 'skunk bear' and 'Indian devil.' Wolverines live in coniferous mountain forests, and their need for pristine wilderness has resulted in a scarce and scattered distribution throughout much of the province, but they may be extending their range back into their former territory.

As with other mustelids, Wolverines have five toes, but the inner toe rarely registers in the print. Though the forefoot registers a small heel pad, the hind foot rarely does. Because of its low, squat shape, the Wolverine leaves a host of erratic typical mustelid trails: an alternating walking pattern, the typical 2x2 loping pattern with its print pairs or the common loping pattern of three- and four-print groups.

Similar Species: Other mustelids (pp. 58–75) make much smaller prints and their track patterns are less erratic. When only four toes register, Wolverine tracks may be mistaken for Grey Wolf (p. 38) or large Domestic Dog (*Canis familiaris*) tracks, but the pad shapes are very different.

fore

Fore Print
Length: 5.3–10 cm (2.1–4 in)
Width: 5.3–8.5 cm (2.1–3.3 in)

Hind Print
Length: 5.3–7.5 cm (2.1–3 in)
Width: 5–7.5 cm (2–3 in)

Straddle
7.5–18 cm (3–7 in)

Stride
Walking: 18–35 cm (7–14 in)
2x2 Loping: 30–130 cm (1–4.3 ft)
Loping: 30–90 cm (1–3 ft)

Size (male>female)
Length with tail:
    85–100 cm (34–40 in)

Weight
1.4–5.5 kg (3–12 lb)

walking   *2x2 loping*

FISHER (Black Cat)
Martes pennanti

This agile hunter is comfortable both on the ground and in the trees of mixed hardwood forests. Its speed and eager hunting antics make for exciting tracking as it races up trees and along the ground in its quest for squirrels. Sometimes, a Fisher's tracks may lead to the remains of an unfortunate Porcupine (p. 82); the Fisher is one of the few predators to kill and eat Porcupines.

Though all five toes may register, the small inner toe frequently does not. Only the forefoot has a small heel pad that can show up in the print. The Fisher occasionally walks, making a direct-registering alternating track pattern, but it more often 2x2 lopes in typical mustelid fashion, leaving angled pairs of prints, the hind foot direct registering on the fore print. Loping is the Fisher's most common gait, producing prints in groups of three or four (see the River Otter, p. 60). The patterns often vary within a short distance. Fisher tracks are not associated with water, so the name 'Fisher' is a misnomer!

Similar Species: Small female Fisher tracks may be confused with male Marten (p. 66) tracks, but a Marten weighs less and leaves shallower prints. Otters have larger hind feet than forefeet. When only four toes register, Fisher prints may look like a Bobcat's (p. 48).

Marten

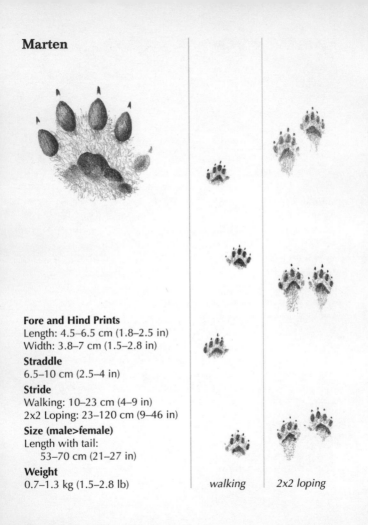

Fore and Hind Prints
Length: 4.5–6.5 cm (1.8–2.5 in)
Width: 3.8–7 cm (1.5–2.8 in)
Straddle
6.5–10 cm (2.5–4 in)
Stride
Walking: 10–23 cm (4–9 in)
2x2 Loping: 23–120 cm (9–46 in)
Size (male>female)
Length with tail:
    53–70 cm (21–27 in)
Weight
0.7–1.3 kg (1.5–2.8 lb)

walking   *2x2 loping*

MARTEN
Martes americana

This aggressive
predator is found
in the coniferous
montane and subalpine
forests of British Columbia.

The Marten seldom leaves a clear print:
the heel pad is undeveloped and, in winter, the
hairiness of the feet often obscures all pad detail, especially the poorly developed palm pads. Often, when the
inner toe fails to register, prints show only four toes. In the
Marten's alternating walking pattern, the hind foot registers
on the fore print. In 2x2 loping, the hind prints fall on the
fore prints to form slightly angled print pairs, in the typical
mustelid pattern. Loping track patterns may appear as
three- or four-print clusters (see the River Otter, p. 60). If
you follow the criss-crossing trails, you may find that the
Marten has scrambled up a tree; look for the sitzmark
where it has jumped down again.

Similar Species: Size and habitat are often key to distinguishing a Marten's tracks from a Fisher's (p. 64) or a
Mink's (p. 68). And, though female Fisher prints overlap in
size with large male Marten prints, Fishers leave clearer
tracks. Similarly, male Mink prints overlap in size with
small female Marten prints, but Minks rarely climb trees
and, unlike Martens, are usually found near water.

67

Mink

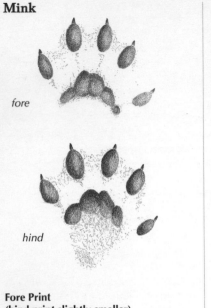

fore

hind

Fore Print
(hind print slightly smaller)
Length: 3.3–5 cm (1.3–2 in)
Width: 3.3–4.5 cm (1.3–1.8 in)

Straddle
5.3–9 cm (2.1–3.5 in)

Stride
Walking/2x2 Loping: 20–90 cm (8–36 in)

Size (male>female)
Length with tail: 48–70 cm (19–27 in)

Weight
0.7–1.6 kg (1.5–3.5 lb)

2x2 loping

MINK
Mustela vison

The lustrous Mink, widespread throughout British Columbia, prefers watery habitats surrounded by brush or forest. At home as much on land as in water, this nocturnal hunter can be an exciting animal to track. Like the River Otter (p. 60), the Mink slides in snow, carving out a trough up to 15 cm (6 in) wide with its body for an observant tracker to spot.

The fore print of the Mink shows five (sometimes four) toes, with five loosely connected palm pads in an arc, but the hind print shows only four palm pads. The metacarpal pad of the forefoot rarely registers, but the furred heel of the hind foot may register, lengthening the hind print. Though its trails show much diversity in gait, the Mink prefers the typical mustelid 2x2 loping, which leaves consistently spaced, slightly angled double prints. Its track patterns also include alternating walking; loping with three- and four-print groups (as illustrated for the River Otter); and bounding, as with rabbits and hares (p. 78).

Similar Species: The weasels (pp. 70–73) make similar trails. Small Martens (p. 66), with similar prints, do not have a consistent 2x2 loping gait or live near water.

Weasels

all weasels

Long-tailed Weasel
Fore and Hind Prints
Length: 2.8–4.5 cm (1.1–1.8 in)
Width: 2–2.5 cm (0.8–1 in)

Straddle
4.5–7 cm (1.8–2.8 in)

Stride
2x2 Loping: 24–110 cm (9.5–42 in)

Size (male>female)
Length with tail: 30–55 cm (12–22 in)

Weight
85–340 g (3–12 oz)

*2x2 loping
(Long-tailed)*

LONG-TAILED WEASEL
Mustela frenata

Weasels are active
year-round hunters,
with an avid appetite for rodents.
Following their tracks can reveal much
about these nimble creatures' activities. Tracks are most
evident in winter, when they frequently burrow into the
snow or pursue rodents into their holes. Some weasel trails
may lead you up a tree. Weasels sometimes take to water.

The typical weasel gait is a 2x2 lope, leaving a trail of
paired prints. Because of a weasel's light weight and small,
hairy feet, the pad detail is often unclear, especially in
snow. Even with clear tracks, the inner (fifth) toe rarely registers. To identify the weasel species, pay close attention to
the straddle and stride, but note that small females of a
larger species and large males of a smaller species may
have similar tracks. Also check the habit displayed in loping patterns and note the distribution and habitat.

The Long-tailed Weasel, found in the south, is the largest of the three in the province, with the largest tracks. Its
typical 2x2 lope shows an irregularity in the length of the
stride—sometimes short and sometimes long—with no consistent behaviour. Like the Mink (p. 68), this weasel may
sometimes bound like a rabbit or hare (p. 78).

Similar Species: A large male Short-tailed Weasel's (p. 72)
tracks may be the same size as those of a small female
Long-tailed Weasel and their ranges do overlap.

Short-tailed Weasel
Fore and Hind Prints
Length: 2–3.3 cm (0.8–1.3 in)
Width: 1.3–1.5 cm (0.5–0.6 in)

Straddle
2.5–5.3 cm (1–2.1 in)

Stride
2x2 Loping: 23–90 cm (9–36 in)

Size (male>female)
Length with tail:
    20–35 cm (8–14 in)

Weight
30–170 g (1–6 oz)

Least Weasel
Fore and Hind Prints
Length: 1.3–2 cm (0.5–0.8 in)
Width: 1–1.3 cm (0.4–0.5 in)

Straddle
2–3.8 cm (0.8–1.5 in)

Stride
2x2 Loping: 13–50 cm (5–20 in)

Size (male>female)
Length with tail:
    17–23 cm (6.5–9 in)

Weight
35–65 g (1.3–2.3 oz)

*2x2 loping
(Short-tailed)*

*2x2 loping
(Least)*

SHORT-TAILED WEASEL
(Ermine, Stoat)
Mustela erminea

Smaller than the Long-tailed Weasel, but larger than the Least Weasel, this weasel is widely distributed. It prefers woodlands and meadows up to high elevations, but does not favour denser coniferous forest or wetlands. Its 2x2 loping tracks may fall in clusters, with alternating short and long strides.

Similar Species: Large male Least Weasel tracks may be the same size as a small female Short-tailed's. A small female Long-tailed's (p. 70) tracks may be the same size as a large male Short-tailed's and ranges do overlap.

LEAST WEASEL
Mustela nivalis

The Least Weasel is the smallest weasel, with the least clear tracks. It inhabits all but the southernmost part of the province. Tracks may be found around wetlands and in open woodlands and fields.

Similar Species: A small female Short-tailed Weasel's tracks may resemble a large male Least Weasel's, but Short-tailed Weasels do not frequent wet areas.

Badger

fore

hind

**Fore Print
(hind print is slightly shorter)**
Length: 6.5–7.5 cm (2.5–3 in)
Width: 5.8–7 cm (2.3–2.8 in)

Straddle
10–18 cm (4–7 in)

Stride
Walking: 15–30 cm (6–12 in)

Size
Length: 53–90 cm (21–36 in)

Weight
6–11 kg (13–25 lb)

walking

BADGER
Taxidea taxus

The squat shape and unmistakable face of this bold animal are most likely to be seen in the open grasslands, but Badgers also venture into high mountain country. They are found throughout southwestern British Columbia. Thick shoulders and fore legs, coupled with long claws, make it a powerful digging animal. Look for Badger tracks in spring and fall snow only—unlike most other mustelids, the Badger likes to den up in a hole for the coldest months of winter.

All five toes on each foot register. A Badger's long claws are evident in the pigeon-toed track that it leaves as it waddles along, although the claws on the hind feet are not as long as on the forefeet. When a Badger walks, the alternating track pattern shows a double register, with the hind print sometimes falling just behind (or sometimes slightly in front of) the fore print. The Badger's wide, low body will often plough through deeper snow, obscuring track detail.

Similar Species: In snow, a Porcupine's (p. 82) trail may be similar, but will show draglines made by its tail and quills, and it will likely lead up a tree, not to a hole.

Striped Skunk

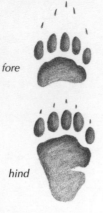

fore

hind

Fore Print
Length: 3.8–5.6 cm (1.5–2.2 in)
Width: 2.5–3.8 cm (1–1.5 in)

Hind Print
Length: 3.8–6.5 cm (1.5–2.5 in)
Width: 2.5–3.8 cm (1–1.5 in)

Straddle
7–11 cm (2.8–4.5 in)

Stride
Walking: 15–23 cm (6–9 in)
Running: 35–50 cm (14–20 in)

Size
Length with tail:
    50–80 cm (20–32 in)

Weight
2.7–6.5 kg (6–14 lb)

walking (fast)   *running*

STRIPED SKUNK
Mephitis mephitis

This striking skunk has a notorious reputation for its vile smell; the lingering odour is often the best sign of its presence. Widespread throughout British Columbia, it prefers lower elevations, in a diversity of habitats. The Striped Skunk dens up in winter, coming out on warmer days and in spring.

Both forefeet and hind feet have five toes. The long claws on the forefeet often register. Smooth palm pads and small heel pads leave surprisingly small prints. Skunks mostly walk—with such a potent smell for their defence, and those memorable black and white stripes, they rarely need to run. A distinctive feature of this skunk (once classed as a mustelid) is that its trail rarely shows any consistent pattern, but an alternating walking pattern may be evident. The greater a skunk's speed, the more the hind foot oversteps the fore. If it runs, its trail consists of clumsy four print groups set closely to each other. In snow it drags its feet.

Similar Species: The Western Spotted Skunk (Civet Cat, *Spilogale gracilis*), in the extreme southwest, has smaller prints, in a very random pattern. Similar-sized Domestic Cat (p. 50) prints in snow may be confused with a skunk's.

77

Snowshoe Hare

fore

hind

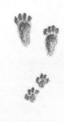

Fore Print
Length: 5–7.5 cm (2–3 in)
Width: 3.8–5 cm (1.5–2 in)
Hind Print
Length: 10–15 cm (4–6 in)
Width: 5–9 cm (2–3.5 in)
Straddle
15–20 cm (6–8 in)
Stride
Hopping: 25–130 cm (10 in–4.3 ft)
Size
Length: 30–53 cm (12–21 in)
Weight
0.9–1.8 kg (2–4 lb)

hopping

SNOWSHOE HARE (Varying Hare)

Lepus americanus

This hare is well known for its colour change, from summer brown to winter white, and for its huge hind feet, which enable it to 'float' on the surface of snow. Widespread throughout British Columbia, it frequents brushy areas in forests, which provide good cover from the Lynx (p. 46) and the Coyote (p. 40), its most likely predators. Hares are most active at night.

As with all rabbits and other hares, the Snowshoe Hare's most common track pattern is a hopping one, with groups of four prints in a triangular pattern. These groups can be quite long if the hare is moving quickly. The hind prints are much larger than the fore prints. In winter, heavy fur on the hind feet thickens the toes, which can splay out to further spread the hare's weight when it runs on snow. Follow a hare's trail and you might notice the well-worn runways, which are often used as escape runs. If you are lucky, you might even come across a resting hare, because hares do not live in burrows. Twigs and stems that have been neatly severed at a 45° angle are a sign of this hare's presence.

Similar Species: Mountain (Nuttall's) Cottontail (*Sylvilagus nuttallii*), with similar but smaller prints, occurs only in a localized region near British Columbia's southern border.

Pika

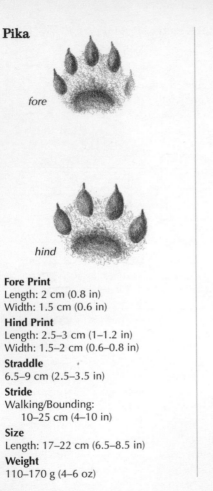

fore

hind

Fore Print
Length: 2 cm (0.8 in)
Width: 1.5 cm (0.6 in)

Hind Print
Length: 2.5–3 cm (1–1.2 in)
Width: 1.5–2 cm (0.6–0.8 in)

Straddle
6.5–9 cm (2.5–3.5 in)

Stride
Walking/Bounding:
10–25 cm (4–10 in)

Size
Length: 17–22 cm (6.5–8.5 in)

Weight
110–170 g (4–6 oz)

bounding

PIKA (Cony, Rock Rabbit)
Ochotona princeps

When hiking high up in the mountains, you are more likely to hear the squeak of this cousin of the rabbit than to see it, because it is quick to disappear under the rocks for protection when alarmed. Confined to areas of high elevation, the Pika rarely leaves good tracks: it prefers exposed rocky areas on mountain slopes; its tracks are most likely to be found in spring, on patches of snow or in mud. A more conspicuous sign of the Pika's presence is its little hay piles, set to dry in the sun for the winter ahead, during which the Pika feeds on its stored food as it remains active under the snow.

The fore print usually shows five toes, unless the fifth toe does not register, but the hind print shows only four. The prints may appear in an erratic alternating pattern or in three- and four-print bounding groups.

Similar Species: The Pika's high-mountain habitat means that its prints are rarely confused with rabbit or hare (p. 78) prints, though track patterns may be similar. At the southern border, Mountain (Nuttall's) Cottontail (*Sylvilagus nuttallii*) fore prints never show five toes, and the hind prints have a long heel.

Porcupine

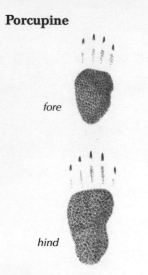

fore

hind

Fore Print
Length: 5.8–8.5 cm (2.3–3.3 in)
Width: 3.3–4.8 cm (1.3–1.9 in)
Hind Print
Length: 7–10 cm (2.8–4 in)
Width: 3.8–5 cm (1.5–2 in)
Straddle
14–23 cm (5.5–9 in)
Stride
Walking: 13–25 cm (5–10 in)
Size
Length with tail: 65–100 cm (25–40 in)
Weight
4.5–13 kg (10–28 lb)

walking

PORCUPINE
Erethizon dorsatum

This notorious rodent
rarely runs, because its many
long quills are a formidable defence.
Widespread throughout British Columbia,
the Porcupine shows a preference for forests,
but it can also be seen in more open areas.

The most common Porcupine track pattern is an alternating walking pattern, with the longer hind foot registering on or slightly in front of the fore print. Look for long claw marks on all prints. The fore print shows four toes and the hind print five. On clear prints, the unusual pebbly surface of the solid heel pads may show, but a Porcupine's pigeon-toed footprints are often obscured by scratches from its heavy, spiny tail. A Porcupine's waddling gait shows in its track pattern. In deeper snow, this squat animal drags its feet, and it may leave a trough with its body. A Porcupine's trail might lead you to a tree, where these animals spend much of their time feeding—look for chewed bark or nipped twigs lying on the forest floor.

Similar Species: The Badger (p. 74) also has pigeon-toed prints, but it does not drag its tail or climb trees.

Mountain Beaver

fore

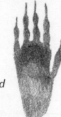

hind

Fore and Hind Prints
Length: 4–5 cm (1.6–2 in)
Width: 2.5 cm (1 in)
Straddle
5–7.5 cm (2–3 in)
Stride
Walking: 7.5 cm (3 in)
Size
Length: 24–48 cm (9.5–19 in)
Weight
0.5–1.4 kg (1–3 lb)

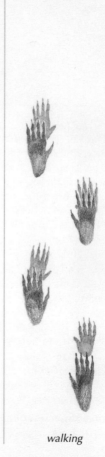

walking

MOUNTAIN BEAVER
(Aplodontia)
Aplodontia rufa

This stocky rodent is a curious resident of the dark, moist forests of southwestern British Columbia. Its tail is so short that it appears not to have one. The Mountain Beaver—neither a beaver nor a montane animal—is an enthusiastic burrower that leaves a labyrinth of shallow tunnels. Its burrowing activity can easily be seen in spring after the snow melts to reveal soil cores much like those of a Northern Pocket Gopher (p. 104), only larger.

Mountain Beavers leave narrow prints, with the hind print showing a slightly longer heel, in an alternating walking track pattern. The hind foot occasionally registers on the fore print. Both forefeet and hind feet have five toes, but the inner toe does not fully register in the fore prints. Mountain Beavers are slow-moving animals that like to stay near home, so you are more likely to notice other signs of Mountain Beaver activity, such as the large hay piles left on logs or on the ground—look for burrows nearby. Trees and shrubs may have shoot ends neatly nipped off; sometimes this animal's tendency to clip leaders and peel bark off conifers can cause serious damage to trees.

Similar Species: Other animals of the dark, moist forest have different prints and activities than this large rodent.

85

Beaver

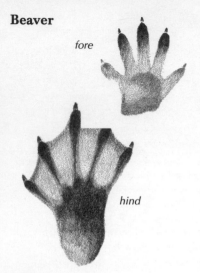

fore

hind

Fore Print
Length: 6.5–10 cm (2.5–4 in)
Width: 5–9 cm (2–3.5 in)
Hind Print
Length: 13–18 cm (5–7 in)
Width: 8.5–13 cm (3.3–5 in)
Straddle
15–28 cm (6–11 in)
Stride
Walking: 7.5–17 cm (3–6.5 in)
Size
Length with tail: 90–120 cm (3–4 ft)
Weight
13–34 kg (28–75 lb)

walking

BEAVER
Castor canadensis

Few animals leave as many signs of their presence as the Beaver, the largest North American rodent and a common sight around waterbodies throughout the province. Common signs include the conspicuous dams and lodges—capable of changing the local landscape—and the stumps of felled trees. Check trunks gnawed clean of bark for marks of the Beaver's huge incisors. A scent mound marked with castoreum, a strong-smelling yellowish fluid that Beavers produce, is also a sign of recent activity.

A Beaver's tracks are often made less clear by its thick, scaly tail or by the branches that it drags about for construction and food. If you find clear tracks, check the large hind prints for signs of webbing and broad toenails—the nail of the second inner toe usually does not show. Rarely do all five toes on each foot register. Irregular foot placement in the alternating walking gait may produce a direct or a double register. Beavers often use the same path repeatedly, resulting in a well-worn trail.

Similar Species: Little confusion arises, because the Beaver leaves many distinctive signs of its presence, including its large footprints.

Muskrat

fore

hind

Fore Print
Length: 2.8–3.8 cm (1.1–1.5 in)
Width: 2.8–3.8 cm (1.1–1.5 in)
Hind Print
Length: 4–8 cm (1.6–3.2 in)
Width: 3.8–5.3 cm (1.5–2.1 in)
Straddle
7.5–13 cm (3–5 in)
Stride
Walking: 7.5–13 cm (3–5 in)
Running: to 30 cm (1 ft)
Size
Length with tail: 40–65 cm (16–25 in)
Weight
0.9–1.8 kg (2–4 lb)

walking

MUSKRAT
Ondatra zibethicus

Like the Beaver, this rodent is found throughout British Columbia, wherever there is water. Beavers (p. 86) are very tolerant of Muskrats and even allow them to live in parts of their lodges. Muskrats are active all year, and they leave plenty of signs of their presence. They dig extensive networks of burrows, often undermining the riverbank, so do not be surprised if you suddenly fall into a hidden hole! Other signs of these rodents are their small lodges in the water and the beds of vegetation on which they rest, sun and feed during summer.

The small inner toe of the five on each forefoot rarely registers. The hind print shows five well formed toes that may have a 'shelf' around them, created by stiff hairs that aid in swimming. The common alternating walking trail shows track pairs that alternate from side to side, with the hind print just behind or slightly overlapping the fore print. In snow, a Muskrat's feet drag and its tail leaves a sweeping dragline.

Similar Species: Few animals share this water-loving rodent's habits. Beavers make larger prints and leave many other signs.

Hoary Marmot

fore

hind

Fore and Hind Prints
Length: 4.5–7 cm (1.8–2.8 in)
Width: 2.5–5 cm (1–2 in)

Straddle
8.5–15 cm (3.3–6 in)

Stride
Walking: 5–15 cm (2–6 in)
Running: 15–35 cm (6–14 in)

Size (male>female)
Length with tail:
    43–80 cm (17–32 in)

Weight
2.3–7 kg (5–15 lb)

walking   *running*

HOARY MARMOT
Marmota caligata

These endearing squirrels seem to lead the good life, sleeping all winter and sunbathing on rocks in summer. They inhabit high rocky areas throughout British Columbia. Hoary Marmots live in small colonies, with an extensive network of burrows, and they are a joy to watch as they play-fight.

The fore print shows four toes and three palm pads; two heel pads may also be evident. The hind print shows five toes, four palm pads and two poorly registering heel pads. In a marmot's usual alternating walking pattern, the hind foot registers over the fore print but, when a marmot runs, it leaves groups of four prints, with the two hind prints in front of the fore prints. Because of its preference for rocky habitats, this marmot's tracks can be hard to find; it is best to look after spring or fall snowfalls.

Similar Species: The rare Vancouver Island Marmot (*Marmota vancouverensis*) makes similar tracks on the Island. The Yellow-bellied Marmot (*Marmota flaviventris*), which lives in central and southern British Columbia, has smaller tracks. Small Raccoons (p. 52) may leave similar running track patterns, but showing five toes on each fore print.

Golden-mantled Ground Squirrel

fore

hind

Fore Print
Length: 2.5–3.3 cm (1–1.3 in)
Width: 1.3–2.5 cm (0.5–1 in)

Hind Print
Length: 2.8–3.8 cm (1.1–1.5 in)
Width: 2–3.3 cm (0.8–1.3 in)

Straddle
5.8–10 cm (2.3–4 in)

Stride
Running: 18–50 cm (7–20 in)

Size
Length with tail:
    20–33 cm (8–13 in)

Weight
170–280 g (6–10 oz)

running

GOLDEN-MANTLED GROUND SQUIRREL

Spermophilus lateralis

The Golden-mantled Ground Squirrel is common throughout central and southern British Columbia. It enjoys varying terrain, from open forests to rocky areas above treeline. These bold squirrels can be very tame when they beg for food, but do not give in, no matter how sweet they are—our unhealthy diet certainly does not suit them!

Ground squirrels hibernate during winter; their tracks may be evident in late or early snowfalls or in mud around their many burrow entrances. The small fifth toe of the forefoot rarely registers in the fore print; the two heel pads sometimes show. The larger hind print shows five toes. Both forefeet and hind feet have long claws that often show in the prints. Ground squirrels are usually seen scurrying around, leaving a typical squirrel track pattern—the hind prints registering ahead of the fore prints, which are usually placed diagonally.

Similar Species: Columbian Ground Squirrels (*S. columbianus*) and Arctic Ground Squirrels (*S. parryii*) are larger. Yellow-pine Chipmunk (p. 94) tracks are smaller. Tree squirrel (pp. 96–99) tracks have a more square-shaped running group.

Yellow-pine Chipmunk

fore

hind

Fore Print
Length: 2–2.5 cm (0.8–1 in)
Width: 1–2 cm (0.4–0.8 in)
Hind Print
Length: 1.8–3.3 cm (0.7–1.3 in)
Width: 1.3–2.3 cm (0.5–0.9 in)
Straddle
5–8 cm (2–3.2 in)
Stride
Running: 18–38 cm (7–15 in)
Size
Length with tail: 18–23 cm (7–9 in)
Weight
30–70 g (1–2.5 oz)

running

YELLOW-PINE CHIPMUNK
Tamias amoenus

This colourful chipmunk is found in various habitats—anywhere from dry sagebrush to coniferous forest—especially where there is yellow pine. You are more likely to see or hear this rodent, which is highly active in summer, than to find its tracks, because it happily climbs trees to extract its favourite seeds from ripe pine cones. Chipmunks hibernate in winter, waking up occasionally to eat. If it is very mild, they may emerge and travel about in the snow.

Chipmunks are so light that their tracks rarely show fine details. The forefeet each have four toes, and the hind feet five. Chipmunks run on their toes, so the two heel pads of the forefeet seldom register; the hind feet have no heel pads. Their erratic track patterns show the hind feet registering in front of the forefeet, like many of their cousins. A chipmunk's trail often leads to extensive burrows.

Similar Species: The Least Chipmunk (*T. minimus*), found in the north, has smaller tracks. Townsend's Chipmunk (*T. townsendii*) inhabits the extreme southwest. The Red-tailed Chipmunk (*T. ruficaudus*) lives in the extreme southeast. Midwinter tracks are more likely a squirrel's (pp. 92–93, 96–99). Red Squirrel tracks are larger and have a larger straddle. Mouse (pp. 108–11) tracks are smaller.

 95

Red Squirrel

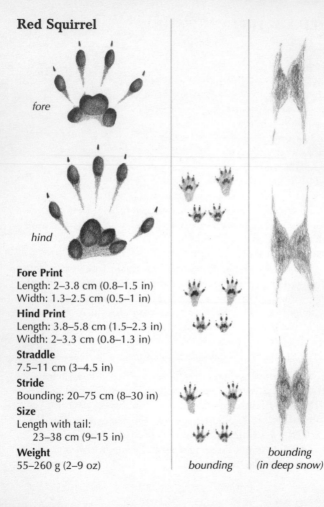

fore

hind

Fore Print
Length: 2–3.8 cm (0.8–1.5 in)
Width: 1.3–2.5 cm (0.5–1 in)

Hind Print
Length: 3.8–5.8 cm (1.5–2.3 in)
Width: 2–3.3 cm (0.8–1.3 in)

Straddle
7.5–11 cm (3–4.5 in)

Stride
Bounding: 20–75 cm (8–30 in)

Size
Length with tail:
    23–38 cm (9–15 in)

Weight
55–260 g (2–9 oz)

bounding

*bounding
(in deep snow)*

RED SQUIRREL
(Pine Squirrel, Chickaree)
Tamiasciurus hudsonicus

When you enter
a Red Squirrel's terri-
tory, the inhabitant
greets you with a loud,
chattering call. Another
obvious sign of this province-
wide forest dweller is its large
middens—piles of cone scales and
cores left beneath trees—that indicate
favourite squirrel feeding sites.

Active year-round in their small territories, Red Squirrels
leave an abundance of trails that lead from tree to tree or
down a burrow. These energetic animals mostly bound.
This gait leaves groups of four prints, with the hind prints
falling in front of the fore prints, which tend to be side by
side (but not always). Four toes show on each fore print,
and five on each hind print. The heels often do not register
when squirrels move quickly. In deeper snow, the prints
merge to form pairs of diamond-shaped tracks.

Similar Species: The larger Fox Squirrel (*Sciurus niger*)
is found in the extreme south of the province. Yellow-pine
Chipmunk (p. 94) and Northern Flying Squirrel (p. 98)
tracks show a similar pattern, but have a smaller straddle
and smaller prints. A flying squirrel's hind feet rarely register in front of its fore tracks.

Northern Flying Squirrel

hind

Fore Print
Length: 1.3–2 cm (0.5–0.8 in)
Width: 1.3 cm (0.5 in)
Hind Print
Length: 3.3–4.5 cm (1.3–1.8 in)
Width: 2 cm (0.8 in)
Straddle
7.5–9.5 cm (3–3.8 in)
Stride
Bounding: 28–75 cm (11–30 in)
Size
Length with tail: 23–30 cm (9–12 in)
Weight
110–180 g (4–6.5 oz)

sitzmark into bounding

NORTHERN FLYING SQUIRREL

Glaucomys sabrinus

This acrobat can be found in coniferous forests throughout all of British Columbia. It prefers widely spaced forests, where it can glide from tree to tree through the night, using the membranous flap between its legs. Northern Flying Squirrels will den up together in a tree cavity for warmth.

Because of its gliding, this squirrel does not leave as many tracks as other squirrels, and it is very difficult to find any evidence of its presence in summer. In winter, however, you might come across a sitzmark—the distinctive pattern that a squirrel leaves when it lands in the snow—and a short bounding trail, made as it rushes off to the nearest tree or to do some quick foraging. The bounding track pattern is typical of squirrels and other rodents, but with the hind feet registering only slightly in front of the fore prints—though all four feet often register in a row.

Similar Species: Red Squirrels (p. 96) usually make larger prints and rarely leave a sitzmark, but unclear tracks in deep snow can be impossible to distinguish from a flying squirrel's. Yellow-pine Chipmunks (p. 94) have a smaller straddle and smaller prints.

Bushy-tailed Woodrat

fore

hind

Fore Print
Length: 1.5–2 cm (0.6–0.8 in)
Width: 1–1.3 cm (0.4–0.5 in)

Hind Print
Length: 2.5–3.8 cm (1–1.5 in)
Width: 1.5–2 cm (0.6–0.8 in)

Straddle
5.8–7 cm (2.3–2.8 in)

Stride
Walking: 4.5–7.5 cm (1.8–3 in)
Bounding: 13–20 cm (5–8 in)

Size
Length with tail:
    28–48 cm (11–19 in)

Weight
200–600 g (7–21 oz)

walking   *bounding*

BUSHY-TAILED WOODRAT
(Packrat)
Neotoma cinerea

The Bushy-tailed Woodrat is found almost everywhere in British Columbia, especially in rugged terrain and forests, but it is not as fond of deserts as other woodrat species are. Tracking one of these nocturnal rodents can be very rewarding: the trail might lead you to its distinctive massed nest, which can be 1.5 m (5 ft) across and may be in an abandoned building—but do not be surprised if its trail ends at the base of a tree. This curious hoarder brings home all manner of objects and serves as a rather selective wilderness garbage collector.

Woodrats often walk in an alternating fashion, with the hind foot direct registering on the fore print. The stride is short, relative to the size of the prints. The fore print shows four toes and the hind print five. A woodrat's short claws rarely register. This woodrat frequently bounds, leaving a track pattern of four prints, with the larger hind feet registering in front of the diagonally placed forefeet.

Similar Species: Other, less abundant, species of woodrat have similar prints. Norway Rats (p. 102) are usually found close to human activity. Hoary Marmot (p. 90) prints are similar, but much larger.

Norway Rat

fore

hind

Fore Print
Length: 1.8–2 cm (0.7–0.8 in)
Width: 1.3–1.8 cm (0.5–0.7 in)

Hind Print
Length: 2.5–3.3 cm (1–1.3 in)
Width: 2–2.5 cm (0.8–1 in)

Straddle
5–7.5 cm (2–3 in)

Stride
Walking: 3.8–9 cm (1.5–3.5 in)
Bounding: 23–50 cm (9–20 in)

Size
Length with tail: 33–48 cm (13–19 in)

Weight
200–510 g (7–18 oz)

walking

NORWAY RAT
(Brown Rat)
Rattus norvegicus

This despised rat is widespread almost anywhere humans have decided to build homes, although it is not entirely dependent on people and may live in the wild as well. It is active both day and night.

The fore print shows four toes and the hind print five. When it bounds, this colonial rat leaves groups of four prints, with the hind prints in front of the diagonally placed fore prints. Sometimes one of the hind feet directly registers on a fore print, creating a track pattern of only three prints. More commonly, rats leave an alternating walking pattern, with the larger hind prints close to or overlapping the fore prints, and the hind heel does not show. In snow, the tail often leaves a dragline. Rats live in groups, so you may find many trails together, often leading to their 5-cm (2-in) wide burrows.

Similar Species: Bushy-tailed Woodrat (p. 100) tracks may be similar, but woodrats rarely associate with human activity, except in abandoned buildings.

Northern Pocket Gopher

fore

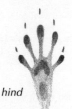

hind

Fore Print
Length: 2.5 cm (1 in)
Width: 1.5 cm (0.6 in)
Hind Print
Length: 2–2.5 cm (0.8–1 in)
Width: 1.3 cm (0.5 in)
Straddle
3.8–5 cm (1.5–2 in)
Stride
Walking: 3.3–5 cm (1.3–2 in)
Size (male>female)
Length with tail: 15–23 cm (6–9 in)
Weight
80–140 g (2.8–5 oz)

walking

NORTHERN POCKET GOPHER
Thomomys talpoides

This seldom-seen rodent is found in southern British Columbia, anywhere from open pine forests to high alpine meadows. It spends most of its time in burrows, only venturing out to move mud around or to find a mate. Because of their need to dig, pocket gophers prefer soft, moist soils.

By far the best sign of Northern Pocket Gopher activity is the muddy mounds and tunnel cores—they are especially evident just after spring thaw. Each mound marks the entrance to a burrow, which is always blocked up with a plug. Search around the mounds to find tracks. Both forefeet and hind feet have five toes. The forefeet have well-developed long claws for digging, though it is a rare track that shows this much detail. Pocket gophers typically leave an alternating walking track pattern, with the hind foot registering on or slightly behind the fore print.

Similar Species: Northern Pocket Gopher tracks are associated with the distinctive burrows, leaving little room for confusion, except with other, less abundant, pocket gopher species.

Meadow Vole

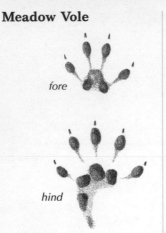

fore

hind

Fore Print
Length: 0.8–1.3 cm (0.3–0.5 in)
Width: 0.8–1.3 cm (0.3–0.5 in)

Hind Print
Length: 0.8–1.5 cm (0.3–0.6 in)
Width: 0.8–1.5 cm (0.3–0.6 in)

Straddle
3.3–5 cm (1.3–2 in)

Stride
Walking/Trotting:
   3.3–7.5 cm (1.3–3 in)
Bounding: 10–20 cm (4–8 in)

Size
Length with tail:
   14–20 cm (5.5–8 in)

Weight
14–170 g (0.5–6 oz)

walking

*bounding
(in snow)*

MEADOW VOLE
(Field Mouse)
Microtus pennsylvanicus

There are so many vole species that positive trail identification is next to impossible. One possible way to differentiate the Meadow Vole from other vole species is to note that it is commonly found in damp or wet habitats across most of the province.

Vole fore prints show four toes and hind prints show five, but they are seldom clear. A vole's walk and trot both leave a paired alternating track, with the hind foot occasionally registering directly on the fore print. Voles usually opt for a faster bounding, however, with the prints appearing in pairs, hind foot registering on fore print. When crossing open areas, these voles lope quickly, creating a three-print track pattern. Voles stay under the snow in winter and so, when the snow melts, look for distinctive piles of cut grass from their ground nests. The bark at the bases of shrubs may show tiny teeth marks left by gnawing. In summer, well-used vole paths appear as little runways in the grass.

Similar Species: The Montane Vole (*M. montanus*) is commonly found at higher elevations in central and southern British Columbia. The Long-tailed Vole (*M. longicaudus*) inhabits drier regions. Townsend's Vole (*M. townsendii*) is found on Vancouver Island. The Deer Mouse (p. 108) makes a track pattern with shorter strides, and its fore prints have only four toes.

Deer Mouse

bounding group

Fore Print
Length: 0.8–1 cm (0.3–0.4 in)
Width: 0.8–1 cm (0.3–0.4 in)
Hind Print
Length: 0.8–1.3 cm (0.3–0.5 in)
Width: 0.8–1 cm (0.3–0.4 in)
Straddle
3.6–4.5 cm (1.4–1.8 in)
Stride
Bounding: 13–30 cm (5–12 in)
Size
Length with tail:
    15–23 cm (6–9 in)
Weight
14–35 g (0.5–1.3 oz)

bounding

bounding
(in snow)

DEER MOUSE
Peromyscus maniculatus

The Deer Mouse is
the most abundant
mammal in the moun-
tains, but it is seldom seen
because it is nocturnal. This
highly adaptable rodent lives anywhere from arid valleys all the way
up to alpine meadows. It may enter
buildings and stay active in winter.

The fore prints each show four toes, three palm pads
and two heel pads. The hind prints show five toes and
three palm pads; the heel pads rarely register. It takes perfect, soft mud to get clear prints from such a tiny mammal.
Bounding tracks, most noticeable in snow, show the hind
prints falling in front of the fore prints. In soft snow, the
prints may merge to appear as larger pairs of prints, with
tail drag evident. Follow the tracks, and they may take you
up a tree or down into a burrow.

Similar Species: Tracks of many less common species of
mice are identical. House Mouse (*Mus musculus*) tracks are
very similar, but they are associated more with humans.
Meadow Voles (p. 106) tend to trot and the bounding track
pattern of voles is much shorter. Yellow-pine Chipmunks
(p. 94) have a wider straddle. Western Jumping Mouse
(p. 110) prints may be similar in size, but have long, thin
toes. Dusky Shrews (p. 112) have a narrower straddle.

Western Jumping Mouse

*bounding
group*

Fore Print
Length: 0.8–1.3 cm (0.3–0.5 in)
Width: 0.8–1.3 cm (0.3–0.5 in)

Hind Print
Length: 1.3–3.3 cm (0.5–1.3 in)
Width: 1.3–1.8 cm (0.5–0.7 in)

Straddle
4.5–4.8 cm (1.8–1.9 in)

Stride
Bounding: 18–45 cm (7–18 in)
In alarm: 90–180 cm (3–6 ft)

Size
Length with tail: 18–23 cm (7–9 in)

Weight
17–35 g (0.6–1.3 oz)

bounding

WESTERN JUMPING MOUSE
Zapus princeps

Congratulations if you find and successfully identify the tracks of a Western Jumping Mouse! Though it is distributed widely in the province, its preference for grassy meadows—and its long, deep winter hibernation (about six months!)—make locating tracks very difficult.

Jumping mouse tracks are distinctive if you do find them. The two smaller forefeet register between the long hind prints; the long heels do not always register and some prints show just the three long middle toes. The toes on the forefeet may splay so much that the side toes point backward. When bounding, these mice make short leaps. The tail may leave a dragline in soft mud or unseasonable snow. Clusters of cut grass stems, about 13 cm (5 in) long, lying in meadows, are a more abundant sign of this rodent.

Similar Species: Pacific Jumping Mouse (*Z. trinotatus*) lives in the extreme southwest of the province. The Deer Mouse (p. 108) has a similar-sized straddle. Jumping mouse tracks may look so abstract that they can be mistaken for those of a small bird or amphibian (pp. 134–37).

Dusky Shrew

bounding group

Fore Print
Length: 0.5 cm (0.2 in)
Width: 0.5 cm (0.2 in)
Hind Print
Length: 1.5 cm (0.6 in)
Width: 0.8 cm (0.3 in)
Straddle
2–3.3 cm (0.8–1.3 in)
Stride
Bounding: 3–5 cm (1.2–2 in)
Size
Length with tail: 7.5–13 cm (3–5 in)
Weight
3–30 g (0.1–1 oz)

bounding

DUSKY SHREW
Sorex monticolus

Though many species of tiny, frenetic shrews are found in British Columbia, the most likely candidate is the widespread and adaptable Dusky Shrew. This shrew is just as happy in the high heathlands as in the lowland swamps. Its rapid activity makes it difficult to observe closely.

In its energetic and unending quest for food, a shrew usually leaves a bounding pattern of four prints, but it may slow to an alternating walking pattern. The individual prints in a group are often indistinct but, in mud or shallow, wet snow, you can even count the five toes on each print. In deeper snow, a shrew's tail often leaves a dragline. If a shrew tunnels under the snow, it may leave a ridge of snow on the surface. You may find that a shrew's trail disappears down a burrow.

Similar Species: The Wandering Shrew (*S. vagrans*) prefers moist habitats. The Water Shrew (*S. palustris*) is larger and is often found near cold mountain streams. The Masked Shrew (*S. cinereus*) is nocturnal and more secretive. Mice (pp. 108–11) show just four toes on each fore print.

Sasquatch

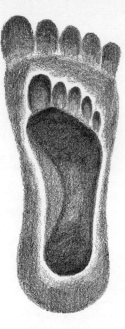

Footprint
Length: 35–43 cm (14–17 in)
Width: to 18 cm (7 in)

Hand Print
Length: to 30 cm (1 ft)
Width: to 18 cm (7 in)

Stride
Walking: 60–90 cm (2–3 ft)

Size (male>female)
Height: 1.8–2.4 m (6–8 ft)

Weight
180–450 kg (400–1000 lb)

SASQUATCH (Bigfoot)

Homo cascadensis

Imagine the thrill of being one of the few people to find footprints of the reclusive Sasquatch!

This elusive, human-like inhabitant of the remote wilderness has an enormous print that is much broader and more flat-footed than a human's (*Homo sapiens*)—notice the raised arch on the human print. Its track patterns are like a human's. Most Sasquatch tracks have been found along rivers, but a clear print might be seen on firmer surfaces. If you find a naked print of huge dimensions, let somebody know!

Similar Species: Huge bear (pp. 34–37) prints may be mistaken for Sasquatch prints, but the many distinctions include prominent claws on the bear prints. Juvenile Sasquatch tracks might be hard to distinguish from large human footprints, but naked human footprints are mostly seen on sandy beaches near water, where (in summer) humans can be seen lying around in large numbers, like lazy sea lions. Elsewhere, the human foot is usually shod to protect it from sticks and stones. Also, irresponsible humans leave plenty of evidence of their presence on trails. If a woodrat has not beaten you to it, do the beautiful mountain wilderness a big favour and pick up the garbage.

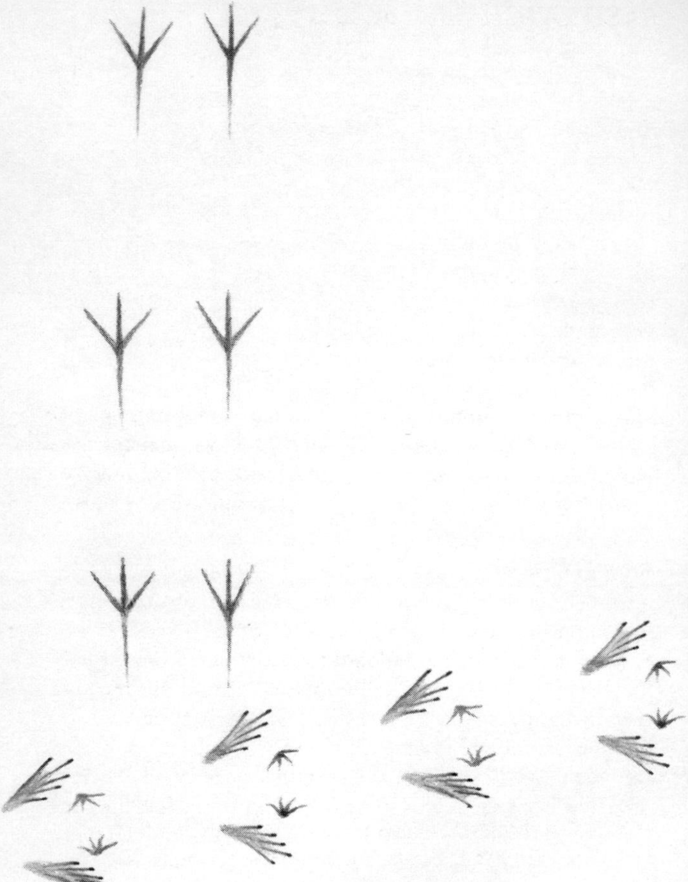

BIRDS, AMPHIBIANS AND REPTILES

A guide to the animal tracks of British Columbia is not complete without some consideration of the province's birds, amphibians and reptiles.

Several bird species have been chosen to represent the main types common to this region, but the differences between bird species are not always reflected in their tracks.

Bird tracks can often be found in abundance in snow and are clearest in shallow, wet snow. The shores of streams and lakes are very reliable locations to find bird tracks—the mud there can hold a clear print for a long time. The sheer number of tracks made by shorebirds and waterfowl can be astonishing. Though some bird species prefer to perch in trees or soar across the sky, it can be entertaining to follow the tracks of those that do spend a lot of time on the ground. They can spin around in circles and lead you in all directions. The trail may suddenly end as the bird takes flight, or it might terminate in a pile of feathers, the bird having fallen victim to a hungry predator.

Many amphibians and turtles depend on moist environments, so look in the soft mud along the shores of lakes and ponds for their distinctive tracks. Though you may be able to distinguish frog tracks from toad tracks, because they generally move differently, it can be very difficult to identify the species. In drier environments, reptiles outnumber amphibians, but dry terrain does not show prints well, except in sand. Snakes can leave distinctive tracks as they wind their way through mud or sand; footless, they make body prints.

117

Canada Goose

Print
Length: 10–13 cm (4–5 in)
Straddle
13–18 cm (5–7 in)
Stride
Walking: 13–18 cm (5–7 in)
Size
80–120 cm (2.7–4 ft)

CANADA GOOSE
Branta canadensis

This common goose is a familiar sight in open areas by lakes and ponds. Its huge, webbed feet leave prints that can often be seen in abundance along the muddy shores of just about any waterbody, including those in urban parks, where the Canada Goose's green-and-white droppings can accumulate in prolific amounts.

The webbed feet each have three long toes, all facing forward. These toes register well, but the webbing between them does not always show on the print. The prints point inward, which gives the bird a pigeon-toed appearance and may account for its waddling gait.

Similar Species: Many waterfowl, such as ducks and gulls, leave similar, but usually smaller, prints. Exceptionally large tracks are likely a swan's.

Herring Gull

Print
Length: 9 cm (3.5 in)
Straddle
10–15 cm (4–6 in)
Stride
11 cm (4.5 in)
Size
Length: 58–65 cm (23–25 in)

HERRING GULL
Larus argentatus

Herring gulls, with their long wings and webbed toes, are strong long-distance fliers as well as excellent swimmers. They are found throughout British Columbia, and are concentrated in great numbers on the coast. Their tracks show three toes, and are slightly asymmetrical. They have claws that register outside the webbing and the claw marks are usually attached to the footprint. Most gulls have quite a swagger to their gait, and they leave a trail with the tracks turned strongly inward.

Similar Species: Gull species cannot be reliably identified by track alone, but smaller species have conspicuously smaller tracks. Duck tracks are difficult to distinguish from gull tracks. Canada Goose (p. 118) and swan tracks are usually larger.

Great Blue Heron

Print
Length: to 17 cm (6.5 in)
Straddle
20 cm (8 in)
Stride
23 cm (9 in)
Size
1.3–1.4 m (4.2–4.5 ft)

GREAT BLUE HERON
Ardea herodias

The refined and graceful image of this large heron symbolizes the precious wetlands in which it patiently hunts for food. Usually still and statuesque as it waits for a meal to swim by, this heron will have cause to walk from time to time, perhaps to find a better hunting location. Look for its large, slender tracks along the banks or mudflats of waterbodies.

Not surprisingly, a bird that lives and hunts with such precision walks in a similar fashion, leaving straight tracks that fall in a nearly straight line. Look for the slender rear toe in the print.

Similar Species: Cranes (*Grus* spp.) often choose similar habitats and have similarly sized prints, but their smaller hind toes do not register.

Spotted Sandpiper

Print
Length: 2–3.3 cm (0.8–1.3 in)
Straddle
to 3.8 cm (1.5 in)
Stride
Erratic
Size
18–20 cm (7–8 in)

SPOTTED SANDPIPER
Actitis macularia

The bobbing tail of the Spotted Sandpiper is a common sight on the shores of lakes, rivers and streams, but you will usually find just one in any given location. Because of its excellent camouflage, likely the first you will see of this bird will be when it flies away, its fluttering wings close to the surface of the water.

As it teeters up and down on shores, it leaves trails of three-toed prints. Its fourth toe is very small and faces off to one side at an angle. Sandpiper tracks can have an erratic stride.

Similar Species: These tracks are quite typical of all sand pipers, plovers and the Common Snipe (*Gallinago gallinago*), although there will be much diversity in size.

Ruffed Grouse

Print
Length: 5–7.5 cm (2–3 in)

Straddle
5–7.5 cm (2–3 in)

Stride
7.5–15 cm (3–6 in)

Size
38–48 cm (15–19 in)

RUFFED GROUSE
Bonasa umbellus

This ground-dweller prefers the quiet seclusion of mixed forests in winter, so that will be the best place to find its tracks. If you follow them quietly, you may be startled when the grouse bursts from cover underneath your feet. Its excellent camouflage usually affords it good protection.

The three thick front toes leave very clear impressions, but the short rear toe, which is angled off to one side, will not always show up so well. This bird's neat, straight trail appears to reflect its cautious approach to life on the forest floor.

Similar Species: The White-tailed Ptarmigan (*Lagopus lecurus*) and the Blue Grouse (*Dendragapus obscurus*) leave similar tracks, but their prints may be enlarged and obscured by the winter feathers that they grow on their feet.

Great Horned Owl

Strike
Width: to 90 m (3 ft)
Size
55 cm (22 in)

GREAT HORNED OWL
Bubo virginianus

This wide-ranging owl is often seen resting quietly in trees during the day, as it prefers to hunt at night. An accomplished hunter in snow, this owl leaves a 'strike' that can be quite a sight if it registers well. The owl strikes through the snow with its talons, leaving an untidy hole, which is occasionally surrounded by imprints of wing and tail feathers. These feather imprints are made as the owl struggles to take off with possibly heavy prey. It is not the most graceful of walkers, preferring to fly away from the scene.

You might stumble across this owl's strike and guess that its target could have been a vole scurrying around underneath the snow. If you are a really lucky tracker, you will follow the surface trail of an animal to find that it abruptly ends with this strike mark, where the animal has been seized by an owl or perhaps a hawk.

Similar Species: Other large, predatory birds also leave strike marks. If the prey left no approaching trail, the strike mark is likely an owl's, because owls hunt by sound. If there is a trail, the strike mark could be from a hawk or a Common Raven (p. 130), both of whom hunt by sight.

 129

Common Raven

Print
Length: to 10 cm (4 in)
Straddle
to 10 cm (4 in)
Stride
Walking: to 15 cm (6 in)
Size
60 cm (2 ft)

COMMON RAVEN
Corvus corax

This legendary bird spends a lot of time strutting around on the ground—confident behaviour that may hint at its intelligence.

Ravens show a typical alternating track pattern. Prints show three thick toes pointing forward and one toe pointing backward. When a Raven is in need of greater speed, perhaps for takeoff, it leaves a trail of diagonally placed pairs of prints that are rather irregular.

Similar Species: Other corvids also spend a lot of time poking around on the ground. Their tracks are similar, but smaller, and their strides are correspondingly shorter. For example, American Crow (*Corvus brachyrhyncos*) prints are up to 7.5 cm (3 in) long and Black-billed Magpie (*Pica pica*) prints are up to 5 cm (2 in) long.

Dark-eyed Junco

Print
Length: to 3.8 cm (1.5 in)
Straddle
2.5–3.8 cm (1–1.5 in)
Stride
Hopping: 3.8–13 cm (1.5–5 in)
Size
14–17 cm (5.5–6.5 in)

DARK-EYED JUNCO
Junco hyemalis

This common small bird typifies the many small hopping birds found in the province. Each foot has three forward-pointing toes and one longer toe at the rear. The best prints are left in snow, although in deep snow the toe detail is lost; the feet may show some dragging between the hops.

A good place to study these types of prints is near a birdfeeder. Watch the birds scurry around as they pick up fallen seeds, and then have a look at the prints that they have left behind. For example, juncos are attracted to seeds that chickadees scatter as they forage for sunflower seeds in the birdfeeder. Also look for tracks under coniferous trees, where juncos feed on fallen seeds in winter.

Similar Species: The size of the toes may indicate what kind of bird you are tracking—larger birds make larger prints. Not all birds are present year-round, so keep in mind the season when tracking. In powdery snow, junco tracks could be mistaken for mouse (pp. 108–11) tracks—follow the trail to see if it disappears down a hole or into thin air.

Toads and Frogs

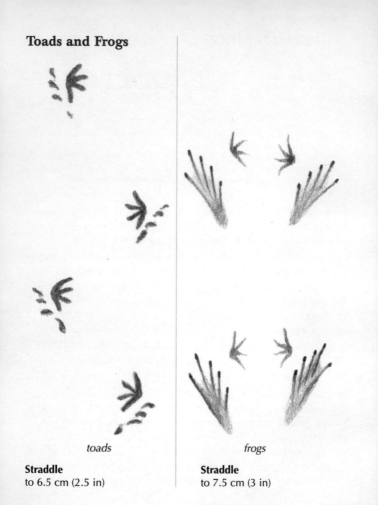

toads

frogs

Straddle
to 6.5 cm (2.5 in)

Straddle
to 7.5 cm (3 in)

TOADS AND FROGS

The best toad and frog tracks are found along the muddy fringes of waterbodies, but toad tracks can occasionally be found in drier areas, perhaps as unclear trails in dusty patches of soil. In general, toads walk and frogs hop, but toads are pretty capable hoppers, too, especially when being hassled by overly enthusiastic naturalists.

TOADS

Widely distributed throughout central and coastal British Columbia, the Western (Boreal) Toad (*Bufo boreas*) inhabits streams, meadows and woodlands. Toads leave rather abstract prints as they walk—the heels of the hind feet do not register. On less firm surfaces, the toes often leave draglines.

Western Toad

FROGS

The small Pacific Treefrog (*Hyla regilla*) lives in southern British Columbia. The much larger Red-legged Frog (*Rana aurora*) is confined to the province's western fringes and Vancouver Island. The hardy, widespread

Wood Frog

Spotted Frog (*R. pretiosa*) frequents cold mountain streams and lakes throughout the province. The Wood Frog (*R. sylvatica*), common even at high elevations, is the most widespread. A frog's hopping action results in the two small forefeet registering in front of the long-toed hind feet. Frog tracks vary greatly in size, depending on species and age.

135

Lizards, Salamanders and Newts

skink walking

Straddle
to 7.5 cm (3 in)

salamander walking

Straddle
to 7.5 cm (3 in)

LIZARDS, SALAMANDERS AND NEWTS

Western Skink

Although some frogs and toads can survive the harsh seasons of northern British Columbia, most amphibian species are unsuited to them. Similarly, it is only in the south that you might find the tracks of the long, slender reptiles known as lizards. Trail identification is difficult, because a dragging belly or a thick tail often smudges the prints. In extreme southern regions, tracks of the Short-horned Lizard (*Phrynosoma douglassi*) can be found on sandy plains or around rocky outcrops in forested areas. More widespread along the southern border, in woodlands and open areas, is the Western Skink (*Eumeces skiltonianus*). The most widespread lizard in southern British Columbia is the Northern Alligator Lizard (*Gerrhonotus coeruleus*), which frequents moist forested areas and lives under rotten logs.

Salamanders in this region are difficult to observe but, with luck, you may find the trail of the Long-toed Salamander (*Ambystoma macrodactylum*) near waterbodies. Another salamander whose trail might be encountered is the Northwestern Salamander (*A. gracile*). Similar to the salamanders is the Rough-skinned Newt (*Taricha granulosa*), a stout-bodied amphibian that leaves comparable, but smaller, prints.

Snakes

typical snake track

SNAKES

Common Garter Snake

There are only a few snakes to be found in British Columbia, with the greatest diversity in the warmer south. Snake trails can be found in mud, dust or loose sand, but identification among the various species is next to impossible, because all snakes are long and slender and make trails that appear so similar. In fact, because a snake lacks feet and leaves a trail that is just a gentle meander, it is very challenging even to establish in which direction the snake was travelling. You can, however, estimate how fast a snake was travelling by the width of its trail. A wide trail with strong side looping indicates that it was moving quickly, whereas a narrow and uneven trail suggests a lower speed.

The Common Garter Snake (*Thamnophis sirtalis*) lives in the southern half of British Columbia. It is often associated with water, in a variety of habitats, including meadows, farmland and valleys. Another small snake in the same region is the Rubber Boa (*Charina bottae*). It does have a rubbery appearance and it frequents damp woodlands. The two largest snakes in the province are the Racer (*Coluber constrictor*), which reaches up to 2 m (6.5 ft) in length, and the Pine-Gopher Snake (*Pituophis melanoleucus*), which can be 2.5 m (8.3 ft) long.

TRACK PATTERNS & PRINTS

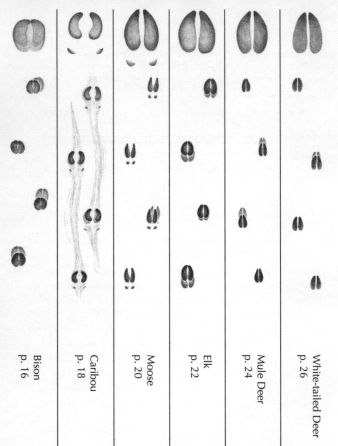

Bison
p. 16

Caribou
p. 18

Moose
p. 20

Elk
p. 22

Mule Deer
p. 24

White-tailed Deer
p. 26

Mountain Goat
p. 28

Bighorn Sheep
p. 30

Horse
p. 32

Grizzly Bear
p. 34

Black Bear
p. 36

Grey Wolf
p. 38

TRACK PATTERNS & PRINTS

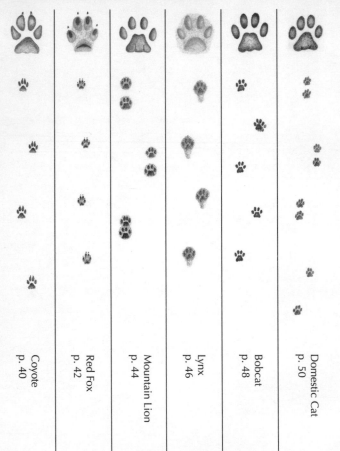

Coyote
p. 40

Red Fox
p. 42

Mountain Lion
p. 44

Lynx
p. 46

Bobcat
p. 48

Domestic Cat
p. 50

142

Raccoon
p. 52

Opossum
p. 54

Harbour Seal
p. 56

Sea Otter
p. 58

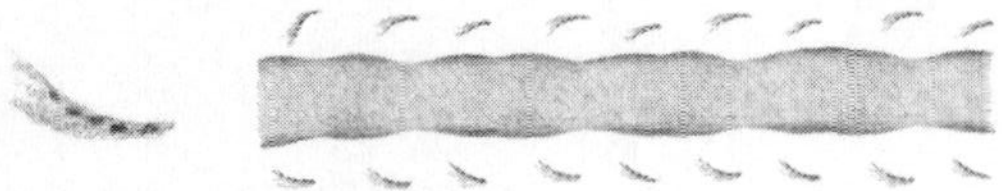

River Otter
p. 60

Wolverine
p. 62

TRACK PATTERNS & PRINTS

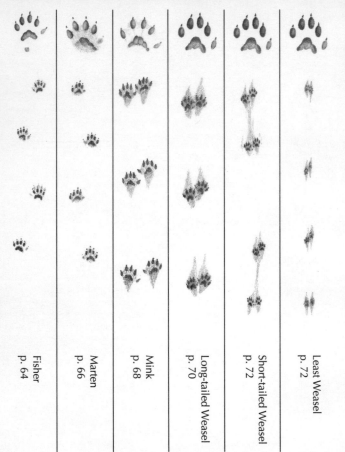

Fisher
p. 64

Marten
p. 66

Mink
p. 68

Long-tailed Weasel
p. 70

Short-tailed Weasel
p. 72

Least Weasel
p. 72

Badger
p. 74

Striped Skunk
p. 76

Snowshoe Hare
p. 78

Pika
p. 80

Porcupine
p. 82

Mountain Beaver
p. 84

TRACK PATTERNS & PRINTS

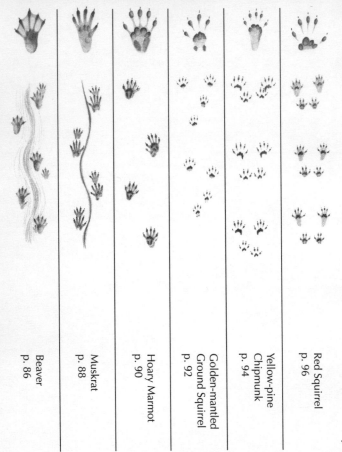

Beaver
p. 86

Muskrat
p. 88

Hoary Marmot
p. 90

Golden-mantled
Ground Squirrel
p. 92

Yellow-pine
Chipmunk
p. 94

Red Squirrel
p. 96

Deer Mouse
p. 108

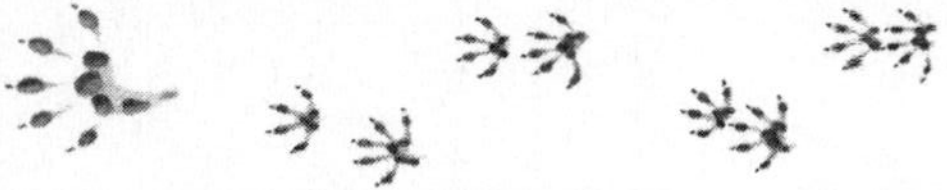

Meadow Vole
p. 106

Northern
Pocket Gopher
p. 104

Norway Rat
p. 102

Bushy-tailed
Woodrat
p. 100

Northern
Flying Squirrel
p. 98

TRACK PATTERNS & PRINTS

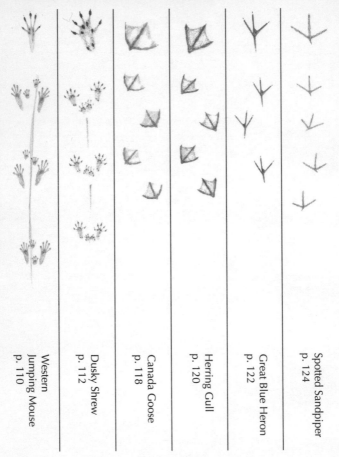

Western
Jumping Mouse
p. 110

Dusky Shrew
p. 112

Canada Goose
p. 118

Herring Gull
p. 120

Great Blue Heron
p. 122

Spotted Sandpiper
p. 124

149

HOOFED PRINTS

Mule Deer

White-tailed Deer

Mountain Goat

Bighorn Sheep

Caribou

Elk

Moose

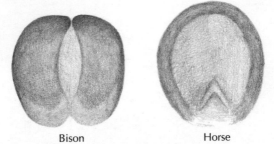

Bison

Horse

inch | cm
0 | 0
1 |
2 | 5

150

FORE PRINTS

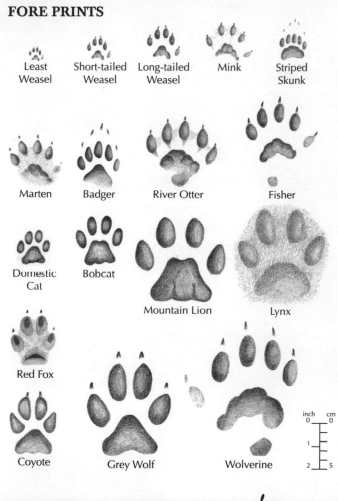

Least Weasel

Short-tailed Weasel

Long-tailed Weasel

Mink

Striped Skunk

Marten

Badger

River Otter

Fisher

Domestic Cat

Bobcat

Mountain Lion

Lynx

Red Fox

Coyote

Grey Wolf

Wolverine

inch cm
0 — 0

1 —

2 — 5

151

FORE PRINTS

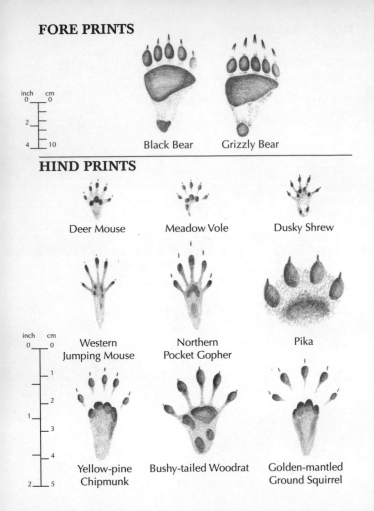

Black Bear

Grizzly Bear

HIND PRINTS

Deer Mouse

Meadow Vole

Dusky Shrew

Western
Jumping Mouse

Northern
Pocket Gopher

Pika

Yellow-pine
Chipmunk

Bushy-tailed Woodrat

Golden-mantled
Ground Squirrel

HIND PRINTS

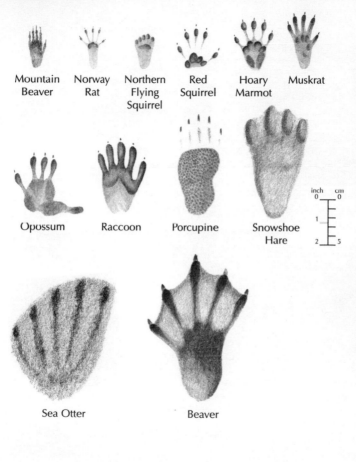

Mountain Beaver

Norway Rat

Northern Flying Squirrel

Red Squirrel

Hoary Marmot

Muskrat

Opossum

Raccoon

Porcupine

Snowshoe Hare

inch cm
0   0

1

2   5

Sea Otter

Beaver

BIBLIOGRAPHY

Barwise, J. E. 1989. *Animal Tracks of Western Canada*. Edmonton: Lone Pine Publishing.

Behler, J. L., and F. W. King. 1979. *Field Guide to North American Reptiles and Amphibians*. National Audubon Society. New York: Alfred A. Knopf.

Burt, W. H. 1976. *A Field Guide to the Mammals*. Boston: Houghton Mifflin Company.

Corkran, C. C., and C. Thoms. 1996. *Amphibians of Oregon, Washington and British Columbia*. Edmonton: Lone Pine Publishing.

Farrand, J., Jr. 1995. *Familiar Animal Tracks of North America*. National Audubon Society Pocket Guide. New York: Alfred A. Knopf.

Forrest, L. R. 1988. *Field Guide to Tracking Animals in Snow*. Harrisburg: Stackpole Books.

Gadd, B. 1995. *Handbook of the Canadian Rockies*. Jasper: Corax Press.

Halfpenny, J. 1986. *A Field Guide to Mammal Tracking in North America*. Boulder: Johnson Publishing Company.

Headstrom, R. 1971. *Identifying Animal Tracks*. Toronto: General Publishing Company.

Kavanagh, J. 1993. *Nature BC*. Edmonton: Lone Pine Publishing.

Murie, O. J. 1974. *A Field Guide to Animal Tracks*. The Peterson Field Guide Series. Boston: Houghton Mifflin Company.

Rezendes, P. 1992. *Tracking & the Art of Seeing: How to Read Animal Tracks & Signs*. Charlotte: Camden House Publishing.

Scotter, G. W., and T. J. Ulrich. 1995. *Mammals of the Canadian Rockies*. Saskatoon: Fifth House.

Stall, C. 1989. *Animal Tracks of the Rocky Mountains*. Seattle: The Mountaineers.

Stokes, D., and L. Stokes. 1986. *A Guide to Animal Tracking and Behaviour*. Toronto: Little, Brown and Company.

Wassink, J. L. 1993. *Mammals of the Central Rockies*. Missoula: Mountain Press Publishing Company.

Whitaker, J. O., Jr. 1996. *National Audubon Society Field Guide to North American Mammals*. New York: Alfred A. Knopf.

INDEX

Page numbers in **boldface** type refer to the primary (illustrated) treatments of animal species and their tracks.

ABOUT THE AUTHORS

Ian Sheldon, an accomplished artist, naturalist and educator, has lived in South Africa, Singapore, Britain and Canada. Caught collecting caterpillars at the age of three, he has been exposed to the beauty and diversity of nature ever since. He was educated at Cambridge University and the University of Alberta. When he is not in the tropics working on conservation projects or immersing himself in our beautiful wilderness, he is sharing his love for nature. Ian enjoys communicating this passion through the visual arts and the written word.

Tamara Hartson, equipped from the age of six with a canoe, a dip net and a note pad, grew up with a fascination for nature and the diversity of life. She has a degree in environmental conservation sciences and has photographed and written about the biodiversity in Bermuda, the Galapagos Islands, the Amazon Basin, China, Tibet, Vietnam, Thailand and Malaysia. When she is not travelling, she enjoys writing and studying biodiversity.